INFORMATION TECHNOLOGY FOR WATER AND WASTEWATER UTILITIES

WEF Manual of Practice No. 33

Second Edition

Edited by Marianne Langridge, PhD

2022

Water Environment Federation
601 Wythe Street
Alexandria, VA 22314–1994 USA
http://www.wef.org

ISBN 978-1-57278-415-4

About WEF

The Water Environment Federation (WEF) is a not-for-profit technical and educational organization of 35,000 individual members and 75 affiliated Member Associations representing water quality professionals around the world. Since 1928, WEF and its members have protected public health and the environment. As a global water sector leader, our mission is to connect water professionals; enrich the expertise of water professionals; increase the awareness of the impact and value of water; and provide a platform for water sector innovation. To learn more, visit www.wef.org.

Second edition edited by Marianne Langridge, PhD

With contributions by:

Daniel E. Capano
David L. Ubert

Review provided by:

Amy Corriveau
Thomas L. Kuczynski
Steve Mustard, PE CEng CAP GISCP CMCP

Text originally prepared by the **Information Technology in Water and Wastewater Utilities Task Force** of the **Water Environment Federation**

Z. Cello Vitasovic, PhD, PE, *Chair*
Marianne L. MacDonald, PhD, *Vice-Chair*

Daniel P. Baker, PE
Michael Barnett, PhD
Hardat Barran, PE, MBA
Burcin Becerik-Gerber, DDes
Rich Castillon
Peter Craan, PE, CAP
Thomas DeLaura
Rienk de Vries, BCE, MBA
George R. Freiber
Donald Gray
Marla K. Hartson, PMP
David Henry, PE
Charles P. McDowell
Jon H. Meyer
Gustaf Olsson, PhD
Steven M. Ravel, PE, BCEE
Melanie Rettie
Randal W. Samstag
Nedeljko Štefanic, Prof. dr.sc.
Michael Waddell
Corey Williams, PE

Under the Direction of the **Automation and Information Technology Subcommittee of the Technical Practice Committee** of the **Technical Practice Committee**

Manuals of Practice of the Water Environment Federation

The WEF Technical Practice Committee (formerly the Committee on Sewage and Industrial Wastes Practice of the Federation of Sewage and Industrial Wastes Associations) was created by the Federation Board of Control on October 11, 1941. The primary function of the Committee is to originate and produce, through appropriate subcommittees, special publications dealing with technical aspects of the broad interests of the Federation. These publications are intended to provide background information through a review of technical practices and detailed procedures that research and experience have shown to be functional and practical.

Water Environment Federation Technical Practice Committee Control Group

Andrew R. Shaw, PhD, PE, BCEE, *Chair*
Jon Davis, *Vice-Chair*
Dan Medina, PhD, PE, *Past Chair*

H. Azam, PhD, PE
G. Baldwin, PE, BCEE
S. Basu, PhD, PE, BCEE, MBA
P. Block, PhD
C.-C. Chang, PhD, PE
R. Chavan, PhD, PE, PMP
A. Deines
M. DeVuono, PE, CPESC, LEED AP BD+C
N. Dons, PE
T. Dupuis, PE
T. Gellner, PE
C. Gish
V. Harshman
G. Heath, PE
M. Hines
M. Johnson
N.J.R. Kraakman, Ir., CEng.
J. Loudon
C. Maher
M. Mulcare
C. Muller, PhD, PE
T. Page-Bottorff
A. Rahman
J. Reina
V. Sundaram, PhD, PE
M. Tam, PE
E. Toot-Levy

Contents

List of Figures

List of Tables

Preface

The purpose of this Manual of Practice (MOP 33) is to provide an overview of information technology (IT) within water and wastewater utilities. Technology and utility's application of technology has evolved significantly since the original publication of MOP 33, yet many of the core principles for the successful application of IT remain. The intention of this new version is to preserve the fundamental principles and most common systems, and to update areas where the application of technology has changed significantly, and to highlight emerging areas and areas where best practices have developed over time, including cybersecurity and data governance.

Perhaps one of the greatest changes in the last decade is that utility staff have come to accept and welcome IT tools as it has become part of the fabric of utility management and operations. Information technology projects still have their share of challenges, but it is clear that these tools are changing the utility business. As new opportunities with artificial intelligence and digital twins emerge, it is important for the stewards of our utilities to keep the fundamentals in mind. This Manual is intended to provide a basic understanding of those fundamentals. The authors of this manual focused on creating a useful and practical summary from an enormous body of information, with the goal of including only content that is most practical to a utility professional.

The second edition of this Manual of Practice was edited under the direction of Marianne Langridge, PhD.

Authors' and reviewers' efforts were supported by the following organizations:

City of Grand Rapids Michigan, Environmental Services Department
District of Columbia Water and Sewer Authority
Electrical Design Associates
Gannett Fleming Engineers and Architects
Hampton Roads Sanitation District (HRSD)
Sustainable Synthesis Limited, PBC
Trinnex, a CDM Smith company
Water and Wastewater CIO Forum

1

Introduction

1.0 EVOLUTION OF INFORMATION TECHNOLOGY

When Manual of Practice (MOP) 33 was first published in 2010, the intent was to capture the standards of practice the had emerged in the application of information technology (IT) in water and wastewater utilities during the previous decades. At that time IT had increasingly been incorporated into all aspects of utility operations, and utilities were making significant investments in IT. Much has changed in the last decade, yet much has stayed the same. At that time it was clear that IT plays a critical role in the management of most utilities, and that the IT field would continue to rapidly change the way we live and work. That continues to be the case today as utilities have come to rely on IT systems to serve their customers, and as they have gained experience with systems and data to address emerging challenges.

Some of the most significant changes in the past decade include

- the move from running critical business systems on premise at a utility to off-site web hosting,

- the increased awareness of the risks associated with IT systems including cybersecurity threats,
- the sophistication of sensors and models to support decision making,
- the recognition and demonstration of the value of utilities' data to manage more effectively,
- experience of public officials in recognizing the value of wastewater and water data to understand public health and inform action,
- the adoption of collaboration tools to support remote work driven by the global pandemic,
- increased strains on water resources due to climate change and emerging contaminants,
- the recognition of social justice issues in the quality and accessibility of our water systems, and
- the need for greater diversity in our workforce.

The water industry's implementation and use of IT systems has evolved in response to these changes, and there is new content in this version to capture practices and tools that utilities have adopted that can be examples for others. Most utilities have at this point adopted core business systems to support customer information and billing, maintenance management, locating assets, finance and accounting, regulatory reporting, human resources, and payroll in addition to the critical operational systems covered in *Automation of Water Resource Recovery Facilities* (MOP 21) (Water Environment Federation [WEF], 2013). An important trend over the past decade has been to bring some of the data from these disparate systems together. System and data integration supports more advanced utility processes including asset management and resiliency planning, yet the potential is still great to do even more. There are many emerging capabilities as the power of data fuels the industry's potential to design, build, and operate our water systems in ways we cannot fully imagine through machine learning, digital twins, and augmented reality.

To ensure that the content of this MOP is useful and relevant, it is important to recognize the greater context of utility operations. While technology evolves at an exponential rate, many of the challenges facing utilities a decade ago remain today, including

- undervaluation of water in the economic markets, resulting in lack of funds to invest in infrastructure and related IT systems;
- lack of resources to drive systemic changes in the use of IT systems; and

- the large gap between the potential of IT and the reality of utilities' ability to fund and adopt new technologies.

2.0 INTENT AND TARGET AUDIENCE

This MOP is intended to provide a resource to bridge the gap between the IT community and water utilities by presenting information about systems and the processes needed to support them. The goal is for readers to better understand what it will take to get the most value from the investment in IT systems. As with the original version, this MOP will not focus on specific software or hardware products but will instead discuss the general IT management practices that stand the test of time a bit more gracefully. In terms of technical issues, this MOP will stay at the conceptual level and frame the material in a context that demonstrates relevance to utility staff and managers, not IT professionals. More specific technical detail may be presented within some of the case studies and in the references provided in each chapter.

Information technology is not an end in itself. It exists to support business functions and provide benefits to the organization and its customers; utilities turn to IT to reduce risks and improve performance, and they expect IT to accomplish specific business objectives. Many business processes within water/wastewater utilities are supported by IT, and some business processes are heavily dependent on IT.

Efforts to implement IT solutions in utilities and to execute IT projects have not gone without their share of challenges and difficulties. In some organizations, IT projects are infamous for being "always late and always over budget"; there may also be a sense that IT does not always fully deliver on its promises. It would be impossible to publish a document that would provide all the information that is required to implement an IT project successfully, to achieve the benefits, and to protect a manager or a technical person in a utility from the risks that are inherent in all aspects of IT (e.g., planning, design, development, procurement, implementation).

This MOP presents an overview of technology that has demonstrated value and provides a reference and a guide that will aid utility managers and staff faced with practical IT issues in their organizations. The MOP is intended to be a document that is broad, addresses most aspects of IT within water and wastewater utilities, and provides some guidance and references to sources that address specific issues in more detail. Each chapter will include a brief overview of the key concepts covered at the beginning and a list of further resources for people interested in learning more about a topic.

Chapter 2 provides an overview of IT within a typical water and/or wastewater utility. It includes brief descriptions of systems and applications and describes a typical "IT landscape" that can be found in a utility. This chapter discusses business drivers for IT in water/wastewater as well as IT applications including business systems, planning systems (e.g., geographic information systems [GIS]), mathematical models, IT systems that support real-time operations (supervisory control and data acquisition [SCADA] and process control), and laboratory information management systems.

Chapter 3 has been added to this MOP to provide more information about the data in the systems discussed in Chapter 2. The value of data as a commodity has grown exponentially in the past decade, and utilities are beginning to recognize the potential the data have to better manage operations and serve customers.

Chapter 4 addresses issues related to planning for IT within the context of a water/wastewater utility. This chapter describes the methodology and practices for developing and implementing a strategic IT plan. This chapter has been modified to include content related to business process management, program planning, and governance that was previously addressed in other chapters.

Chapter 5 presents critical organizational factors that influence IT projects. It reviews roles and responsibilities, governance and organizational culture, and workforce trends.

Chapter 6 addresses issues related to the management of IT projects. Because water/wastewater utility operators must build, operate, and maintain a large physical infrastructure, the utility approach to project management has typically been shaped by extensive experience with traditional "brick-and-mortar" engineering projects. Information technology projects demand different methodologies and approaches; Chapter 6 addresses project management methodologies that are specifically designed for IT.

Chapter 7 provides a more detailed description of specific components that are part of the IT project and its infrastructure. This section has been updated to address the changes in standards of practice in deploying systems.

Chapter 8 has been entirely rewritten to more thoroughly cover cybersecurity challenges facing utilities. This area is complex and dynamic yet critical for all utility management and staff to understand. Because IT systems enable or actually control critical aspects of a utility's business functions, it is important to understand how to manage the vulnerability of IT systems, and this requires all system users' attention and diligence.

Finally, Chapter 9 presents examples and case studies to illustrate the concepts discussed in the MOP.

3.0 UTILITY STRUCTURE AND INFORMATION TECHNOLOGY

There are more than 16 000 publicly owned treatment works and more than 49 000 community water systems in the United States (U.S. Environmental Protection Agency, 2006, 2012). There are many variations in size, governance, customer population, and demographics across all these organizations that influence their ability to invest in and deploy information technology.

The authors of this MOP have tried to present a comprehensive description of different IT systems, issues, and levels of complexity. To achieve this, the MOP contains some content that is primarily applicable to larger utilities that have more complex IT environments. Such large IT environments have more components and a larger IT "footprint." Thus, one can see a more complete picture of how IT could be structured into a comprehensive solution that addresses issues related to managing a water/wastewater utility. Large utilities typically have more elaborate IT environments, bigger projects, more staff, larger budgets, and more complex systems. To some readers with a small-utility perspective, it may appear that there is a bias toward larger utilities. This was not the intent of the authors; however, it was not possible to create an MOP in which all content was applicable to all utilities.

Size is not the sole determining factor in properly structured IT solutions. For example, stricter regulatory requirements might push even a smaller utility toward automation and IT. Similarly, utilities that are embedded in a city, town, or county governing structure can be significantly influenced by the standards, expectations, and resources associated with that structure. Sometimes this can help speed the adoption of certain technologies, and in other cases it can inhibit the utility's ability to invest in IT tools that meet its needs. The best approach for a specific utility will need to take into account the specific business needs that the utility must address. For that reason, it was the intent of the authors to make readers aware of the breadth of issues related to IT at utilities, recognizing that implementation of all the concepts and functionality described will not necessarily be applicable to all utilities.

To apply the material from this MOP to small utilities, the reader may need to "scale down" consideration of some of the content to better address their specific issues. The material presented here is focused on the business needs of a utility and the role of IT in supporting those needs. So while the scale of needs for a large utility will warrant more staff and resources, the basic functionality of IT systems is common across all but the smallest systems. Even a smaller system will need to select solutions that are based on user requirements, and users across all utilities have similar requirements such as generating and tracking work orders. Therefore, the generic information presented in this MOP can be useful, even though it may not always be directly applicable to smaller utilities.

4.0 OPPORTUNITIES

Much has been learned over the past decade about the use of IT in a utility, and from these lessons we can see opportunities. Most of the premises discussed in the original version hold true today, including the need to have utility staff actively involved at all stages of an IT project, the importance of tight collaboration between IT professionals and utility staff, the value of executive-level support, and the importance of having a strategic plan for IT investments that is aligned with the utility's business needs.

In February 2020, a workshop led by Baywork at the Utility Management Conference in Anaheim, California, addressed the needs to ensure that utilities have a digitally capable workforce. The day before this workshop, WEF hosted a Knowledge Development Forum on Intelligent Water Technology Solutions. The KDF included thoughtful presentations on exciting developments in the application of technology in water utilities by thought leaders from industry and academia. The Baywork workshop was led by utilities for utilities and provided a critical lens on the realities facing their use of technology. The sentiment at the workshop was that fundamental challenges persist in using IT to support the day-to-day business of a utility and the continual need for investment and training puts a strain on already scarce resources.

This gap between technological advances and the reality of IT use in utilities must be addressed to bring the value of IT to water service delivery. There are great opportunities to simplify and democratize access to information to provide meaningful insights and guide proper stewardship of water resources. When the original MOP 33 was written, most information system projects required a significant investment of time and money, and these systems were modeled after business software and hardware that was designed for private business and large governmental organizations. There were few systems designed for utilities by people who understand utility operations and management.

There are more tools than ever that are entering the market designed with the utility worker in mind. The move to virtualization has provided utilities that could not previously afford to implement IT tools with access by eliminating the need for on-premise servers and computers. The proliferation of mobile device capabilities has similarly made accessing IT tools and data much easier for utility workers who are not at a desk most of the day. With this, however, come additional cyber risks. To safely take advantage of these opportunities, it is necessary to understand the fundamentals and the business context for IT in a utility. It is the intention of this updated MOP to provide information to demystify IT and the activities needed for a utility to continue its digital journey.

Although the challenges facing utilities may seem daunting at times, there is an opportunity to recognize the potential for collaboration. Utilities, even privately operated ones, are influenced by broader trends in government. This includes the changes to remote work driven by the COVID-19 pandemic, the recognition of the evolving role data in government service deliver, social justice, and public trust. By stepping back, we are reminded that all aspects of society and government are experiencing similar trends. We can then recognize the power of combining resources, forces, and attention across governmental and community organizations, academic institutions, and private businesses. The fact is that all individuals and organizations are stakeholders for water and wastewater utilities, and expanding the perspective by teaming with others could bring additional resources and appreciation of the collective value of technology modernization.

5.0 REFERENCES

U.S. Environmental Protection Agency. (2006). *Community water system survey.* https://nepis.epa.gov/Exe/ZyPDF.cgi?Dockey=P1009JJI.txt

U.S. Environmental Protection Agency. (2012). *Clean watersheds needs survey.* https://www.epa.gov/cwns

6.0 SUGGESTED READINGS

American Water Works Association Research Foundation. (1997). *The utility business architecture: Designing for change.* AWWARF 90726.

Baywork. (2021, May). *The digital worker: Using digital tools to deliver water services* [White paper]. https://baywork.org/wp-content/uploads/2021/06/Baywork_DigitalWorker_Final.pdf

Deloitte Center for Government Insights. (2021). *Government trends 2021: Global transformative trends in the public sector.* https://www2.deloitte.com/us/en/insights/industry/public-sector/government-trends.html?id=us:2el:3dp:di7070:eng:cgi:030421:fg

Drucker, P. F., Garvin, D., Leonard, D., Straus, S., & Brown, J. S. (1998). *Harvard business review on knowledge management.* Harvard Business School Press.

Hammer, M. (1997). *Beyond reengineering: How the process-centered organization is changing our work and our lives.* HarperBusiness.

Porter, M. E. (1985). *Competitive advantage: Creating and sustaining superior performance.* Free Press.

Water Environment Federation. (2013). *Automation of water resource recovery facilities* (4th ed., Manual of Practice No. 21).

Water Environment Federation. (2016). *Intelligent water compilation.*

2

Where Is the Value? Understanding the Business Context for Information Technology

1.0 SUMMARY OF KEY THINGS TO KNOW

The intent of this chapter is to provide an overview of some of the most common applications of information technology (IT) in water and wastewater utilities to help utility professionals understand the business value and fundamentals of different systems. This includes the typical end users, data, outputs, and basic functionality for each system.

- Each IT system exists to support a vital function within the utility, and collectively they provide data to support decision making and action taking.
- Business systems support finance, accounting, procurement, human resources, customer service, knowledge management, and collaboration.
- Utility management systems support asset management, regulatory requirements, project/program management, capital, and construction projects.
- Operations systems include those that support the operation of the physical facilities and infrastructure.
- Planning systems include models of the infrastructure analytics to support decision making.

2.0 BUSINESS CONTEXT

Utilities must continually balance their investments in IT with those to expand and maintain other aspects of the system infrastructure. There is a distinction between operational technology (OT) and IT. The Water Environment Federation's (WEF, 2013) Manual of Practice (MOP) 21, *Automation of Water Resource Recovery Facilities*, addresses the operational technology systems and uses in detail. Here, we will explain the IT systems that support and complement operations to serve customers, and the drivers for continual investment in IT.

For many years, there have been claims that IT creates business efficiencies to enable utilities to do more with less, yet this is rarely measured and documented. For those who have measured it, the time frame for return on investment has often been longer than anticipated. The majority of utilities have at this point adopted IT systems to support their business and have come to accept, if not welcome, the need for IT in the utility. Given the investments made over the past 20 years, IT is now at a point where incremental investments could provide greater returns. Here, we will focus on the drivers to continue to invest in and evolve the use of IT.

2.1 Customer Service

Perhaps the biggest driver for the evolution of IT in a utility is the need to meet the expectations of customers. This includes direct customer interactions, transparency on water quality and regulatory compliance, billing, system reliability, and locational inquiries. The public is technically sophisticated and expects the organizations they do business with to have complete and easy access to customer information. This will continue as the use of mobile devices fuels expectations for information available at one's fingertips, and as the public becomes more engaged in water-related issues including water quality and climate change effects on water resources. This will drive the current trend to deliver information via mobile applications and to integrate data sets to address public questions and concerns.

2.2 Accessing Information

As data sets are created to support utility management, it is easier than ever to access that information from anywhere. Originally the value of this was to support work in the field, but during the pandemic of 2020/2021 it became vital for all staff to have remote access to information to keep utilities running when staff were working remotely. Increasing access to information for broader use increases the value of the data and improves the value of IT across the organization. Collaboration technology greatly improves the handling of routine jobs and reduces administration burdens. Coupled with updated business processes that address information collection, mobile and radio-frequency identification (RFID), and geographic information system (GIS) data, IT can deliver up-to-the-minute visibility into the placement, use, and condition of assets in the field. This will also drive the expansion of virtual and augmented reality to visualize information and collaborate more meaningfully on complex infrastructure projects. This not only provides added flexibility in meeting and designing solutions, but also will enable participants to experience the information about their systems in a way not previously possible.

2.3 Resilience

Damage to, or destruction of, the nation's water supply and water quality infrastructure by human-made or natural disasters could disrupt the delivery of vital human services and threaten public health and the environment.

These will be exacerbated as a result of the effects of climate change. Investment in data and technology security are among the important considerations for today's utilities to mitigate and respond to these risks. Specific examples include cybersecurity, early warning systems with real-time monitoring sensors, epidemiological monitoring of wastewater during public health crises, and emergency response systems, customized to immediately alert key personnel regarding any problems.

2.4 Governmental Incentives and Requirements

Replacing or upgrading water and wastewater systems, meeting rising demand, and accommodating water quality standards is estimated to require an investment of $109 billion per year for the next 20 years (Value of Water Campaign & American Society of Civil Engineers [ASCE], 2020). Federal, state, and municipal agencies are recognizing the importance of reliable water systems on the security and prosperity of the country, including equitable access to clean water across racial and socioeconomic communities. Government is taking an increased role in driving infrastructure modernization, and as a result, new funding programs and regulatory reporting requirements are emerging. Information technology solutions that help utilities with timely and accurate periodic reporting and compliance monitoring could enhance a utility's ability to receive funding and reduce the burden of reporting expectations.

2.5 Sustainability and Social Justice

Global climate change affects drinking water quantity and quality around the world. Global warming may adversely affect water distribution, availability, and quality. Current approaches to resource management are often unsustainable as judged by ecological, economic, and social criteria. As burdens on water resources increase, so does the awareness of the interrelationships between water, energy, social justice, and quality of life. Utilities are recognizing their important role in ensuring quality of life in their communities and will continue to look for new ways to reduce effects on the environment. Information technology plays an important role in working with end users to facilitate the timely collection and use of data needed to calculate and track sustainability- and social justice–related metrics. This could include initiatives such as the development of a "dashboard" to provide progress on key metrics, as well as advanced modeling to support programs and policy initiatives to address community needs.

2.6 Aging Workforce and Changing Demographics

Utilities are in the midst of a significant demographic change as staff retire and a new generation of employees enters the workforce with high technology expectations. Technology implementation and use may help utilities to further implement technical solutions to address staffing issues, enhance work quality and efficiency, and project the image of a high-tech workforce to attract a prospective, talented, and qualified young workforce. Information technology can facilitate knowledge transfer from senior to junior staff. This is especially important for retaining knowledge and insights about the physical assets of a system when insights about location, age, and performance may be in the memories of longtime staff.

Innovation, teamwork, and professional development can be encouraged and facilitated through the use of IT. Leveraging the knowledge and experiences of existing staff to participate in initiatives to streamline and simplify workforce processes can be mutually beneficial for the utility and employees alike, with the proper outreach and incentives. For example, use of technology in the field to capture information as well as to provide important references to as-built conditions on demand can enhance worker safety and reduce work downtime. Another example of the application of technology to support the most effective involvement of workers is through workforce rollout based on skills and availability and innovative employee life-cycle management that aligns employee talents with corporate goals. This will help revitalize the knowledge base being lost as older workers retire.

2.7 Transparency

As a result of the increased speed of communication and availability of data and technology, changes in the global economy and climate, growing populations, and increased infrastructure demand, water and wastewater utilities are required to improve operational and business practices. Solutions connecting project and work management capabilities with scheduling, outage management, and construction planning and solutions integrating core back-office functions and information on customers, meters, hydrants, financials, and the workforce are becoming standard practice to sustain operations and keep public trust. Integrated financial information using standardized processes, providing visibility on capital, operational, and third-party expenditures; monitoring project costs and regulatory risks; and integrating business performance information with management processes are all being achieved by the use of IT; this will continue to expand.

3.0 OVERVIEW OF INFORMATION TECHNOLOGY SYSTEMS FOR UTILITIES

All the systems discussed in this section are important to running a utility. But how do you know your utility's specific needs for IT systems and applications? The answer is that your IT needs are directly related to your business needs.

To determine what, if any, changes need to be made to your current operations, you must first embark on a needs assessment, which is part of strategic planning covered in Chapter 4 of this manual. Depending on the current suite of systems already in use at a utility, the strategic IT planning process may be applied broadly to the entire utility and potential software systems or to a subset of business processes and specific software systems for those processes.

In evaluating and choosing software, there are a number of common terms associated with the type of tool. These are *legacy*, *custom*, and *commercial* systems; *application service provider* (ASP); and *software as a service* (SaaS) options. These terms can be applied to any of the systems discussed in this chapter.

Legacy is used to describe a software system that is older and is typically based on a technology platform that is no longer supported or that was developed with an outdated programming language. This type of system may be highly successful in meeting the business needs of a utility; however, the system poses a business risk in that if it fails it may not be economically or technically feasible to correct the problem, thereby resulting in a gap in functionality. The risk to the utility depends on the type of system and its role in supporting the business.

Custom is a term used to describe an application that is built specifically for an organization based on its functional needs. This approach is typically taken when no commercial product is available to meet business needs. Although custom systems are appropriate in some circumstances, they can be expensive to develop and risky to maintain over the lifetime of the tool.

Commercial off-the-shelf (COTS) is a term for software that is developed and sold as a product by a company that is in the software business. All the systems discussed in this chapter are available as COTS solutions. Each software company designs, develops, and maintains a tool or set of tools based on its interpretation of the functional needs of the market. Most vendors also support user groups and hold user conferences to gather feedback from their customers on business needs and future software enhancements. There are a variety of different pricing models used by software vendors. These include a subscription fee per named user or per concurrent user, and fees based on the volume of data used or number of transactions. There can also be a maintenance fee that is charged that is typically a percentage of the initial software purchase price. This fee includes software maintenance and upgrades as well as end-user technical support.

Commercial off-the-shelf software may also be available from the vendor through an ASP or SaaS option. Through a web-based delivery mechanism, often called "cloud computing," it is possible to access data and applications stored on remote hardware by way of the internet instead of keeping it all in a local workstation. The benefits of such delivery are more options in the end-user access device, such as a smartphone or tablet. The user can be anywhere, and so can the source for data and applications.

The cloud delivery approach adds to the flexibility to scale bandwidth up or down at will as well as the affordability of pay-as-you-go service and subtracts energy-devouring hardware from the local environment. Considerations for such a mechanism include trust in your selected vendor to ensure security and business continuity in the event of a system failure. Selection of this approach involves a contractual arrangement with the software vendor in which the vendor provides the hardware and infrastructure to host the software as well as all the services for software maintenance. This may be a cost-effective option for many utilities, especially those without an IT department or with limited resources to provide software and architecture support. Fees for this service are typically paid monthly or annually based on the number of end users.

Software, itself, is only one element of a software project. Proper use and acceptance of the tool by staff is as important, if not more important. WEF's Utility Analysis and Improvement Methodology (UAIM) and Water Intrapreneurs for Successful Enterprises (WISE) program have demonstrated how articulating a utility's business processes, and understanding the people, process and technology interactions can lead to greater value creation in strategic projects including IT projects. Figure 2.1 shows the three interrelated elements for value creation in a software project: business processes,

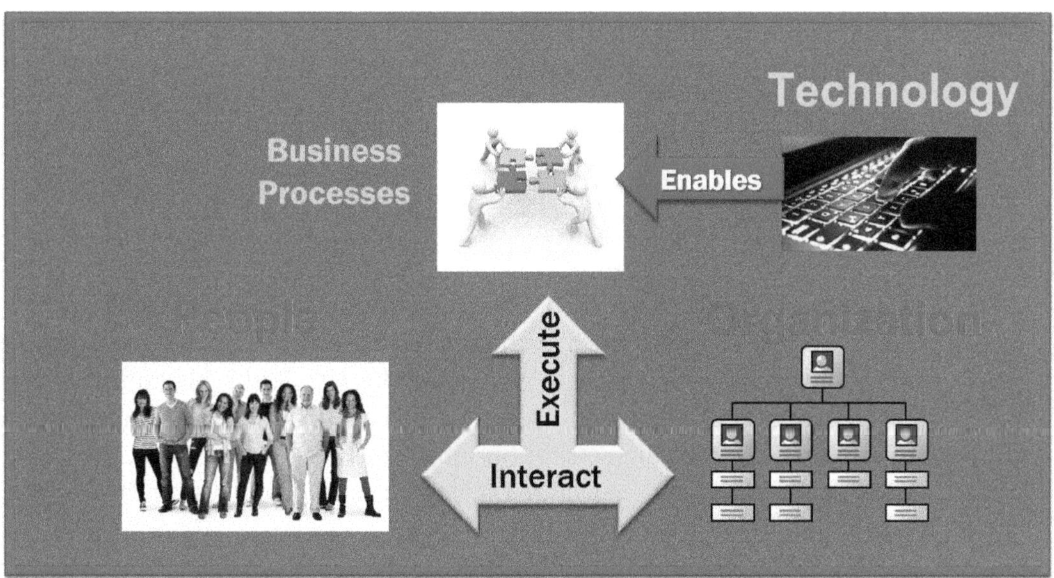

FIGURE 2.1 UAIM Value Creation (Vitasovic et al., 2022) (Reprinted with permission from WEF and IWA Publishing)

technology, and people. It has already been noted that technology should be selected to fit specific business process needs; this topic is explored in greater detail in subsequent chapters. The "people" aspects of technology implementation include considerations such as staff involvement in the definition of requirements and software selection but must carry through to implementation and ongoing use. Training is typically one of the most underfunded and underrecognized critical success factors of any software implementation. Although most COTS vendors provide training on the mechanics of using their software, additional training is often beneficial to educate end users on the role technology plays in performing their business processes, especially if these processes will change because of the introduction of a new software or technology tool. End-user and software support documentation is also critical to successfully maintain software over time, to provide a way for staff to explore lesser used functionality, and to provide information to new staff that come on as end users after the initial implementation.

The following sections provide information on a variety of different, commonly used software tools for a utility. Information provided includes a summary of the defining functions of the tool, typical end users, data inputs, outputs, integration areas, and maintenance considerations. No two systems are identical. Indeed, not only are there differences between custom systems and COTS, but also between different COTS solutions and different implementations of the same vendor COTS solution. These differences occur because each utility has a somewhat different set of needs, business processes, and end users. This should not be considered a problem. Rather, it is because of market demand that vendors enable their software tools to be configured to meet specific organizational needs. Subsequent chapters of this manual will go into more detail about how to plan for, specify, select, and implement the tools needed by your utility.

4.0 BUSINESS SYSTEMS

This section will describe core business functions in utilities and the IT systems that are used to support them. Although there are exciting emerging capabilities in all areas, the focus here will be on the standard tools and uses. For each business area, there is a description of standard IT software application capabilities, the typical users by business function, and the associated data and other information relevant to the implementation and use of the systems.

4.1 Running the Utility Business

As with any business, a utility has employees, customers, and a product, and therefore needs systems to support tracking, managing, and accounting for those things. The effective utility management framework is a useful methodology to assess and understand the components necessary for a well-run utility that serves its community effectively. Establishing and monitoring metrics are a key aspect of this framework, and utility IT systems are necessary to do this.

One of the primary suites of tools used to support utility business management is an enterprise resource planning (ERP) system. An ERP is an information system with multiple modules of functionality serving all the needs of a business including accounting (e.g., general ledger, purchasing, accounts payable, and accounts receivable), human resources, and payroll functionality with configurable workflow and reporting capabilities. Extending ERP can also include the integration of additional data, workflow, and functions such as asset management, work orders, and inventory as well as modules that address customer service, billing, and project management. Additional details can be found in the literature, including *Technologies for Government Transformation: ERP Systems and Beyond* (Kavanagh & Miranda, 2005).

Enterprise resource planning is a widely used approach to organizational information management and operations. An alternate approach is to address a business's functional needs with a suite of distinct software systems that are synchronized or interfaced with each other as necessary, either directly through the software or through business processes. The idea behind an enterprise approach is to enhance operational efficiency by providing management and staff with a unified interface for entering, updating, and accessing information necessary to run a business. For utilities in which IT systems are coordinated with or managed by a municipality, there is additional complexity because of the need to address the needs of multiple departments with different business needs and priorities.

Implementing and updating an ERP system is a significant undertaking. For the software modules to enable a seamless flow of information across an organization, it is necessary to design business processes from human resources to work management and financial management to ensure that roles, responsibilities, and workflow are in sync with the software's intent. Business process mapping is a required element of an ERP implementation and can be used to highlight the role of technology in work execution, including data ownership to ensure that data are entered by the right person at the right time to ensure data quality. Business process mapping is discussed in more detail in Chapter 4.

It is through business process mapping that the specific users of the ERP are defined. Typically, ERP systems are used by staff responsible for accounting and finance, procurement, and purchasing. Depending on the extent of the functionality, it could also include staff involved in payroll, human resources, work management, billing, and asset management. The details on these functions are discussed further in this chapter.

An ERP system includes data related to all financial transactions of a utility because this is the system of record for financial reporting and auditing. When a utility provides financial documents to support grant funding, a bond initiative, and rate changes, this system is used to generate the data and reports needed. It tracks all revenue and expenditures, assets, materials inventory, and depreciation schedules. Some of these items could be obtained through integrations with other systems including human resource, customer information, and asset management systems, or these functions can be addressed as a module within a single ERP.

Challenges associated with ERP implementation are primarily related to the complexity of defining system workflows to accommodate staff in distinct business areas. Figure 2.2 provides an example of how business process mapping can identify systems and users to proactively identify the context for system use. As system functions and data extend to connect staff across functions, an individual may be required to conduct their work differently to better support staff in other business areas. Integration of software functionality through the modules of an ERP requires an integration of, and appreciation for, a broader set of business processes that can be challenging for utility staff. The success of an ERP approach depends, to a great extent, on the culture of the organization and the willingness to commit to a rigorous process of self-assessment and change management.

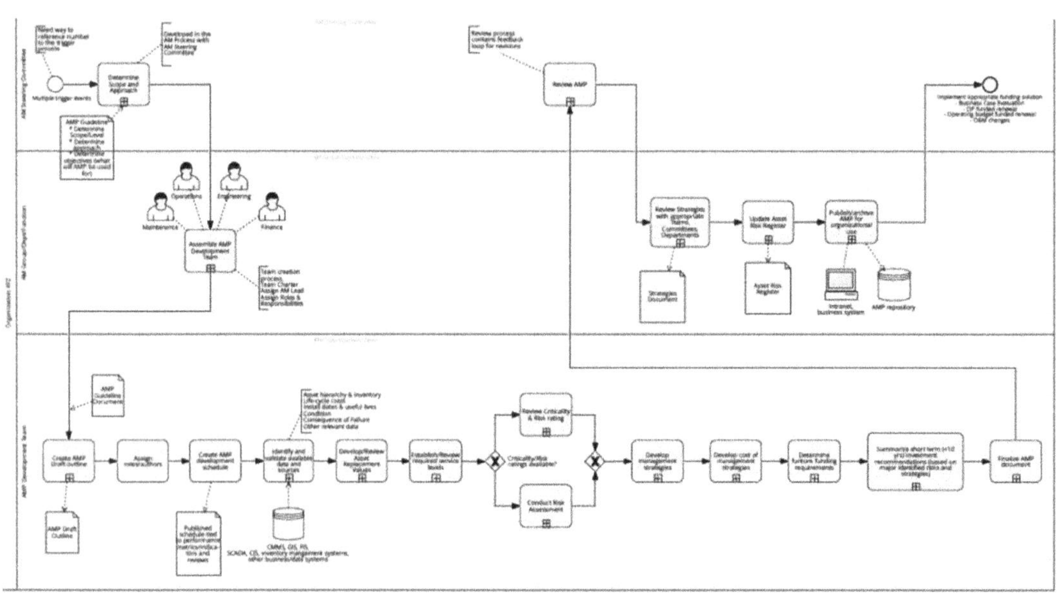

FIGURE 2.2 UAIM/WISE Process Modeling (Vitasovic et al., 2021)

The greatest potential benefit of an ERP approach for a utility may be in the area of asset management. Data on all of the organization's assets, such as people, inventory, and cash flow, are housed in a single database in a manner in which the data can be used in many ways to support consistency in the variety of decisions that must be made. If an organization is committed to the duration and effort required for ERP implementation, the benefits over the long term for the utility can be significant. However, it is possible for an organization to meet its needs and operate efficiently without an ERP if the proper processes and roles are in place.

4.2 Managing and Paying Staff

Staff are the heart of a utility's operation, and IT is needed to support hiring, payroll, work/time recording, benefits administration, training records management, and performance tracking. This functionality may be handled through modules of an ERP system or as a stand-alone human resources (HR) system.

Payroll and employee data management are primary functions of a human resource system. Payroll calculations are conducted by using employee data about salary and benefits with data about time worked each pay period, which results in the issuance of paychecks or direct deposits to staff. "Time-capture" systems may be part of a human resource system and are designed to capture data about time worked through a time entry interface, or they may be a distinct software tool for timekeeping that is integrated with payroll and human resource functionality in another system. Another source of data for labor hours spent on different activities and projects may be a work management system or ERP module. These systems are also configured to track and administer employee benefits such as insurance, Social Security, vacation and sick time, and union participation. The level of sophistication of the functionality will vary by software vendor. All ERP vendors offer human resource functionality, and there are also vendors that specialize in human resource functionality.

System end users are typically limited to designated HR staff and managers because of the privacy requirements of staff's personal data. Human resources end users include the human resource director and any staff responsible for supporting payroll and benefits administration. Some systems may enable direct access to an individual's personal human resource record through the system by that individual and/or the employee's manager.

The core of the human resource system is an employee database. This database may include current and historical information about the employee's skills, degrees, training, and roles in the organization as well as salary and benefits. Other inputs include data about time worked and union participation.

Outputs of the human resource system include payroll for the organization as well as data necessary to support benefits administration such as health insurance and retirement funding. Such systems also allow managers to handle reporting to external organizations such as the Occupational Safety and Health Administration; unions; and Social Security Administration.

Human resource systems may be integrated with timekeeping and work management systems to capture data about time worked. This can facilitate activity-based costing and capital project planning. In addition, they are often integrated with the financial accounting system as employee payroll and benefits are a significant component of cost for the utility. The human resource system can also be integrated with other utility information systems to drive the system security of those systems. For example, roles defined in the human resource system can be used to determine who should have access to other systems, including project management, financial accounting, and customer information systems, where the level of access to the system depends on an employee's role in the organization.

Because of the high potential for multiple integrations with other systems' functionality, software maintenance may be significant if a human resource system is not run as part of an ERP solution. Data maintenance should be built into daily human resource business processes so the timely update of employee information does not become a difficult or time-consuming task. If a human resource system is integrated with other business systems, it is important that there are strong data quality controls and well-defined data ownership responsibilities to maintain data consistency and integrity for employee data across multiple systems.

4.3 Finance and Accounting

Finance and accounting systems focus on the financial management and reporting aspects of the utility. These systems center around financial transactions that are reconciled and reported regularly (typically monthly) to balance income received against payments made. The heart of a financial system is the general ledger and the accounts that transactions are posted against. Typical transactions include budgeting, purchasing, accounts payable, and accounts receivable. When utilities undergo their financial audits annually, this is the system upon which auditors primarily focus.

Direct users of the financial system are the finance department or business operations staff of the utility. This includes staff with responsibility for accounts payable; accounts receivable; and reporting of revenue, expenses, capital assets, and depreciation. Indirect users include all consumers of the utility's financial information such as senior management, board members, public utility commissioners, auditors, and financial institutions.

Data entered into the system include all budgets and financial transactions and the accounts to which they are posted. When a financial system is set up, the structure of the general ledger and the organization of various accounts and account types for the utility are entered. At the beginning of the fiscal year, budgets are entered for all accounts. After this period, financial transactions are entered to the system continuously as they occur.

The outputs of the financial system are reports and statements. These are used by utility management to make business decisions about potential changes required in revenue and expenditures, cash flow, borrowing, and capital planning. In addition, outputs are used for inventory reconciliation, annual financial reporting, and documentation to support changes to billing rates.

The utility's financial system has potential integration points with all the business systems in a utility, including

- inventory—purchasing and accounts receivable,
- customer information systems—revenue from billing and accounts receivable from collections,
- work orders—work costing,
- project management—capital project budgeting,
- asset management—asset valuation, and
- document management.

As with the human resource system, there is a high potential for multiple overlaps with other systems' functionality as it relates to financial data in those systems. Data maintenance may be significant if a financial system is not run as part of an ERP solution. Data maintenance should be built into daily business processes so that the timely update of financial and cost data does not become a difficult or time-consuming task. Reconciliation between an inventory system and a financial system is a common example of areas where maintenance time is required to keep data in sync between systems. If the system is directly integrated with other business systems and not as part of an ERP solution, there may be maintenance issues to address if software upgrades are required by one of the integrated systems. This could have implications for the configuration and upgrade of the financial system.

A utility may choose to replace a financial system because of obsolescence, modernization efforts, or requirements for additional features. Criteria to consider in selecting a new system include required integration points, sophistication of reporting, and ease of use. Data migration between systems as part of an implementation can be straightforward if the originating system is well managed but can become more complicated if integration points are added and additional general ledger/accounts/features are added that affect the configuration.

4.4 Customer Service

Faster response times to customer complaints and claims, improved customer service, round-the-clock self-service, and improved reliability are among the demands of water and wastewater customers. IT-enabled customer service can help utilities meet these demands by making data more easily accessible to support customer inquiries and needs; IT can also help utility staff with easy and fast access to timely accurate customer information for quality decision making. These systems can provide data to support strategic and capital planning and monitor levels of service. This can then lead to increased customer satisfaction and call center efficiency using an integrated customer service collaboration solution that automates standard services and provides fast access and a single view into all data and work history related to customer locations, including water consumption, quality, service disruption, speed of call resolution, and volume of calls.

Customer information systems (CIS) are the systems that enable all this as they house all information about an organization's customers. Customer information systems are primarily used by the customer service and billing staff to support customer inquiries, manage customer accounts and rates, and enable billing and collections. Customer information systems also include historical account information for customers such as payment and usage history; systems can be accessed directly by customers themselves to look up their account history and usage, and to pay bills online.

A key integration of a CIS is with meter reading data. Some systems also include meter inventories and are able to issue service orders for meter and field crews related to customer accounts. When a customer calls to open or close an account or to make an inquiry, the CIS is the primary source of information needed to support that call. The CIS is especially important to the financial operations of a utility because it is the system that manages rates and generates bills and accounts for billing revenues and delinquencies.

Customer service department staff is the primary user of a CIS in terms of inputting data and making inquiries from the system. The customer relationship management functionality of a CIS is quite useful in supporting a customer-oriented organization by tracking customer complaints and inquiries so that the utility can respond in a timely and appropriate manner. If service order functionality is used in the CIS, the meter department and field crews may also use the CIS to receive and close service orders. If the CIS is set up as a web-based system to support direct customer inquiries and payments, the actual customers may also be end users with access to only their personal account.

The database of a CIS centers on customers and locations. Each customer is assigned to at least one location. In addition to location, data typically associated with a customer are payment and usage information, including customer rate type, special payment arrangements, and special conditions that may need to be taken into account, such as medical conditions that prohibit shutting off service and so on. Location information may include an address, property type, assigned meters, and meter location. Customers and locations are typically updated manually as changes occur. Meter data are typically uploaded according to the customer's billing cycle via an upload from a handheld meter reading device or a mobile work management system (service order); from automated meter reading and/or advanced metering infrastructure systems; or sent in from a customer via email, webpage, and/or voice recognition system. The CIS will use these data to calculate usage and resulting bill amounts. Additional details regarding meter reading can be found in the American Water Works Association's *M6: Water Meters—Selecting, Installation, Testing & Maintenance* (2012).

A key output of a CIS is customer bill amounts, which are typically generated on a monthly or quarterly basis for each customer. For most utilities, customers are batched into billing routes and cycles that cover most of a month; billing outputs are generated daily. In addition to bills, CIS generates reports related to billing such as high/low meter readings that warrant investigation and past-due accounts that require notification or collection action. If a customer is past due and a decision is being made about whether to shut off service, set up a payment plan, or issue a warning, the business processes associated with making that decision are all supported by CIS.

Other uses of a CIS may be to provide reports of customers to support notification in the event of a service interruption or water quality event. Customer information system data are also used in a number of business operations and planning areas. These include rate studies, water demand calculations, water loss calculations, and water resource planning.

A CIS is a common component of an ERP because of the many overlaps in data between CIS and financials. When not part of an ERP, the CIS is most often integrated with a meter reading system and the financial system. The points of integration with different systems and the CIS are noted in Table 2.1.

Because the CIS system is the heart of the income stream for a utility, it is important that it be technically reliable and that the quality of data be maintained. Data maintenance of a CIS should be done daily. As customer service staff interact with customers, their account data are typically entered directly to the CIS. It is also useful to periodically run data quality reports

TABLE 2.1 Points of System Integration for a CIS

System	Integration Areas	Purpose of Integration
Meter reading	Meter IDs, meter readings	Provide data for billing
Financial system	Billed revenue, accounts receivable	Cash management and financial planning
GIS	Customer locations	Support routing and customer inquiries

to check customer and location data and to set aside time to make necessary corrections. Ongoing data quality checks can greatly improve data quality and make data conversions much easier and faster if there is a migration to a new system.

Replacing a CIS is a significant undertaking that requires dedicated staff time for planning and implementation. For organizations that are running custom legacy systems, the cost of maintaining those systems and providing the level of service expected from customer rate payers often drives the decision to replace the system with a COTS product. When this is done, it is also a good time to assess the organization's meter reading and financial systems as well to ensure that business processes around the aforementioned integration areas are considered in the selection. For organizations using COTS products, a decision may be made to change vendors if the software is no longer supported or if it is found to be inadequate in addressing current functional requirements of the business.

In addition to the utility's CIS, utilities are increasingly using social media to connect with their customers. Applications such as Facebook and Twitter have the ability to get the word out to customers and the public quickly when necessary, and it is a way customers can let the utility know of conditions they are seeing in the field. Social media outlets also provide a way to build a greater connection to the customer to reinforce the fact that the utility is part of the local community.

4.5 Managing Knowledge and Collaboration

Documentation about utility assets, projects, policies, and processes is critical to ensuring continuity of operations. Systems that support the management of these digital assets may be referred to as *enterprise content management* (ECM) systems, knowledge management (KM) systems, and/or collaboration systems. This is a dynamic area in technology today and one whose value became obvious during the pandemic of 2020/2021 when staff found they needed remote access to information and collaboration tools. These systems capture, index, store, preserve, and deliver digital documents or files for an organization and can enable collaboration in the use of those documents as well.

There are several components to the functionality needed to support knowledge management and collaboration, including

- document/file capture,
- file indexing,
- file retrieval,
- archiving,
- collaboration (such as groupware, instant messaging, and video conferencing),
- web-content management, and
- workflow/business process management (for standardizing and automating business processes).

Once the data is captured in digital form, it will be indexed and categorized for ongoing reference. Data files can be of any type including documents, images, videos, as-built records, multidimensional models. A robust system can handle any type of electronic file.

Typical functions include file search; check-in and checkout; version management; file indexing and navigation; visualization, and organization of data through virtual files, folders, and views. By using web-content management, end users can share information securely over the internet.

Different kinds of ECM/KM repositories (file systems, content management systems, databases, and data warehouses) can be used in combination for storage. Enterprise content management systems also offer long-term, secure electronic archival of static information. There are various long-term storage media such as storage networks; write once, read many, or WORM, optical disks; tapes; and hard disks.

End users of ECM/KM systems can include all utility staff. Security and role-based access can be built into these systems, and employees of a utility and middle and top management typically would have access to the system as viewers. Typically, view-only users have access to the information based on their roles and have no deleting or editing rights. System administrators are a designated group within an organization with authority to enter new information to the system, organize it, and have rights to edit or delete it. Who should have access to what and who can do what should be agreed on and implemented properly as part of the security planning stage of the project.

All documents relevant to an organization should be captured and stored at a centralized ECM/KM system. These include operating procedures, all kinds of reports, manuals, schedules, assessments, standardized forms, plans, drawings, models, surveys, complaints, disputes, pictures, videos, records, and metadata to retrieve and organize data and documents. These inputs are gathered through external sources through manual entry or scanning, database integration with ERP- or customer relationship management (CRM)-types of systems, intranet data, email, and so forth.

Outputs of an ECM/KM system are typically files, documents, and reports that can be shared in various formats such as image file formats (TIFF, JPEG, etc.), PDF, HTML, XML (eXtensible Markup Language), or any other format that is not editable. This information is shared through various distribution channels such as the web or email, both in digital or hard copy format.

Depending on the organization's needs, enterprise and content management systems could be integrated with (a) an ERP or CRM systems such as SAP or PeopleSoft (Oracle Corporation) for workflow management (where ERP or CRM systems support workflow automation and files are stored in an ECM system); (b) collaboration systems such as email clients, chats, or other forms of solutions that support unstructured contents; and (c) portals such as intranets, extranets, or websites for making the portal attractive and more useful. Integration of these systems across the utility's entire IT infrastructure is important to enable end users to access information in the context of their daily workflows. For example, the ability to have a single login window for access to multiple applications reduces a user's need to log into multiple systems to complete a task; however, this can become more technically challenging and costly than it looks. It is important to understand the business processes for which these systems are used to identify where integrations are most needed.

Managing content is one of the main challenges of implementing an ECM/KM system because of the range of people at every level of an organization who produce and access content. One of the most important items that must be addressed in an implementation is the definition of the taxonomy to be used. Taxonomies provide the classification structure to enable end users to easily access content stored in the ECM/KM system. According to IBM, "Having consistent and reliable access to unstructured content is arguably the foundation to realizing the business benefits of ECM, and all subsequent content-centric enterprise applications will realize their return on investment (ROI) by leveraging this essential capability. As large enterprises standardize on ECM platforms, maintaining the integrity of the catalog is essential to managing access to the volume and heterogeneity of business information" (Twigg et al., 2007). Some ECM/KM vendors provide tools and services to support the taxonomy identification. It is important that the utility have a good understanding of its content and access needs. During implementation, the taxonomy will be used to configure the tool and guide the migration of content from existing sources to the new tool. For the tool's use and value to be sustainable for the organization, a governance process must also be established to ensure that the taxonomy is applied; otherwise, content retrieval will not meet initial expectations. Migrating existing information to an ECM/KM system requires a big investment in time and labor, and the volume of the content might become overwhelming as formatting the original document, identifying metadata for each document, and importing them to the ECM/KM system require following a systematic strategic process. In addition, getting approval for implementing an ECM/KM system can be challenging as the cost of implementing and maintaining these systems typically are not trivial. Although the "people factor" is not related to software and data maintenance considerations, it is important to mention here. Often, end users would be accustomed to storing their files in local hard drives and sharing the content company-wide might become an obstacle as it requires a culture change. With the version control capabilities of an ECM/KM, content creators should be open to making documents more widely accessible to support the benefits of information sharing and collaboration for their team and for the broader organization without worrying about document integrity. In addition, supporting and maintaining an ECM/KM system adds responsibilities to the IT department of an organization. There are web-based systems that could reduce the day-to-day burden on IT, and this decision should be considered as part of an overall IT support strategy as discussed in Chapter 4.

5.0 UTILITY MANAGEMENT SYSTEMS

In addition to the business functions discussed above, the utility's primary mission is to deliver water of an acceptable quality to the communities it serves. There are many business processes necessary to fulfill this mission that will be discussed in this section.

5.1 Metering

Measuring water consumption is a fundamental requirement for a water utility to collect revenue from its customers. Accurate and timely meter reading data is necessary not just for sound financial management, but also to provide early warning of anomalies that might indicate a leak. Many utilities collect data using automated meter reading technologies that collect readings for upload to the billing system to support billing. There are also a number of advanced meter technologies that are used to monitor water flows and quality to enable timely response to wet weather and potential contamination. Additional details about metering can be found in the Suggested Readings section below.

5.2 Maintenance and Asset Management

Utilities are responsible for the assets necessary to collect, treat, and distribute or discharge water. Asset management has matured considerably since the first version of this MOP was published. At that time, the focus was on tracking maintenance activities using a computerized maintenance management system (CMMS). Functionally, maintenance and asset management are different sets of processes that coincide to support utility management and capital planning. The CMMS was initially developed to replace paper-based maintenance logs and schedules in the facility and in the field. As utilities realized the importance and value of cataloging physical assets, system capabilities were expanded to catalog assets with their characteristics like condition and maintenance history and assess them geographically and financially to support capital planning and repair and renewal strategies.

Enterprise asset management (EAM) provides a framework for a utility's maintenance and asset management activities. Figure 2.3 presents the basic IT components of an EAM program. Typical functionality includes issuing work orders, supporting preventive maintenance, evaluating condition, enabling job cost accounting, controlling inventory, and facilitating EAM. As with all IT systems, EAM has web-enabled functionality to support remote work and mobile business processes. The core functional areas are described briefly in the following sections.

End users of an EAM span the organization and are based on the work activity being performed. Data entry is best handled by those staff with the most direct knowledge of the information being captured. For example, planners and schedulers will enter data necessary to plan and prepare for the work, and work crews will document work performed. If a large-scale asset data population effort is being done, a coordinated effort between those staff with the greatest knowledge of the data should be undertaken.

Consumers of information in the system include operations and maintenance staff receiving work orders or inquiring about asset conditions for work that they are going to complete. Additional end users also include customer service and accounting staff and senior management responsible for short- and long-term capital planning, budgeting, and risk management. A comprehensive EAM with good data will ultimately result in more coordinated and educated capital decisions.

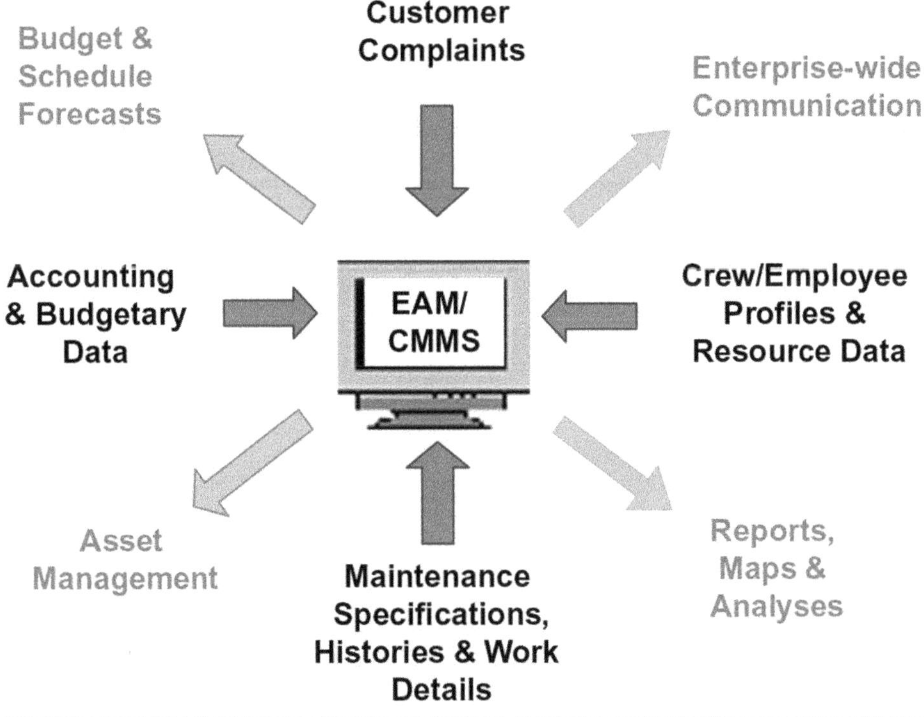

FIGURE 2.3 Role of EAM/CMMS in Utility Operations

Additional activities and decisions supported by the outputs of an EAM include responding to inquiries about asset locations, work history, and asset conditions. In planning and scheduling work, a system that is integrated with a spatial analysis tool like a GIS will enable geo-locating of maintenance work. At the managerial level, the system will support the development and tracking of key performance indicators that allow for a comparison of productivity statistics both internally and externally.

To achieve the maximum value from an EAM solution, there are a number of systems that can be integrated. However, as previously mentioned, it is not the technical system integration that drives the value, it is the coordination of business processes and synchronization of data between systems that ensures the value is achieved. Nontechnical integration with the system listed in Table 2.2 can provide value, even when technical integration is too costly.

As with many of the systems discussed in this chapter, maintenance costs must be considered and planned for in undertaking technological integrations between systems. If one system or module of functionality is replaced with a new system or undergoes a significant upgrade, it can have ripple effects on the other integrated systems. For an EAM, there is the additional consideration of identifying the system of record for asset data. This may be

TABLE 2.2 Points of System Integration for an EAM/CMMS

System	Integration Areas	Purpose of Integration
GIS	Asset and customer locations	Routing and work scheduling Customer inquiries Geographic trend analysis
Financial system	Asset (value), maintenance budget/costs, and capital budget/costs	Governmental Accounting Standards Board reporting Capital planning Job costing for reimbursement Inventory purchasing
CIS	Customer locations	Link service calls to work orders Call logging
Human resources	Employee data	Job costing of labor resources Work assignment based on staff skills
Operations and maintenance manuals	Equipment servicing guidelines/standards	Preventive maintenance scheduling
Process control	Equipment run times	Preventive maintenance scheduling

the EAM, the GIS, or the financial system, depending on the asset management approach of the utility. Regardless of the system of record for asset data, it is critical to have accurate and timely data on assets for the proper operation of a utility. Data entry and data maintenance responsibilities must be clearly defined and regularly completed to support the potentially powerful capabilities of an EAM. For example, establishing standard operating procedures and policies for access to operating information, such as process and instrumentation diagram drawings, can ensure the data consistency and quality that is necessary for an EAM system to have maximum value for an organization.

5.2.1 Asset Management

The EAM is the core tool for asset management in its ability to track an asset's life-cycle costs (labor and material) and rehabilitations. When new assets are built, it is possible to use the as-built building information management (BIM) and virtual design and construction (VDC) files to establish the new assets in the EAM. Most assets, however, were put in place before the availability of this digital data; this requires a significant effort to catalog existing assets for the EAM. Once the data are available, however, the EAM can be a valuable tool in supporting optimization of repair versus replacement decisions as well as the development of a capital improvement plan that is based on asset condition and risk.

The implementation of an EAM alone does not result in an asset management program. The general stages of an asset management program include

- developing an asset register by organizing assets in hierarchy and capturing basic asset attributes,
- determining the relative criticality of the asset,
- assessing the condition of the assets and estimating the remaining useful life,
- assessing the asset risk,
- developing a maintenance strategy consistent with risk,
- developing a replacement plan consistent with risk, and
- developing replacement schedules.

Enterprise asset management tools enable capturing, storing, evaluating, and reporting on the data necessary to make decisions and take action to optimize the life of a utility's assets. The data include asset classifications, location and descriptive information about condition, and history. There are a wide variety of commercial tools on the market that vary in terms of complexity and scalability of the data collected. More sophisticated EAM tools may be part of an ERP strategy that includes multiple modules with data and functionality that includes all the aforementioned functionality as well as fleet management, resource management, and facilities management. Additional references address the topic of implementing asset management programs in more detail (American Water Works Association Research Foundation, 2006, 2008).

5.2.2 *Maintenance*

Maintaining assets within a facility or across a pipe network requires significant effort to prevent system failures. This may include scheduled cleanings or repair, or emergency activities as a result of a break. Keeping a regular maintenance schedule and recording the work completed helps ensure that the utility runs as designed and can serve the public's expectations. The EAM includes electronic records for planned and completed maintenance. Before completing the work, the system can provide staff with immediate access to asset information and maintenance histories. As work is completed, data about maintenance activities, materials, and labor used are entered to the system to support other business needs. By compiling this information, the EAM can then support the analysis of trends in maintenance activities, clustering of issues, and, potentially, cause and effect of activities and outcomes. Linking EAM to GIS can provide even greater functionality in finding work locations, developing routes and work schedules, and conducting trend analyses.

An EAM can help move utilities from a reactive response to issues and problems to proactive management of assets and infrastructure. The system is populated with the recommended maintenance schedule for key pieces of equipment and can then automatically generate preventive maintenance work orders based on a defined schedule. This schedule may be based on run times or calendar time. As work is done, it can then track preventive maintenance activities, ultimately providing data necessary to understand trends in asset conditions. The overall result is the minimization of equipment downtime.

5.2.3 Inventory Management

An EAM can also manage inventory and relate that inventory to work orders. Functionality may include tracking requisitions, automating reordering by setting thresholds and triggers for minimum on-hand quantities, and tracking unit price and average price as well as equipment and material usage on work orders. This information can be quite valuable in providing information to support budgeting for operations and capital planning.

5.2.4 Job Cost Accounting

It is often desirable to be able to identify the total cost of completing a "job" or set of work activities. This may be conducted for accounting reasons as well as to pursue reimbursement from an outside party for work done. The EAM uses work order information to aggregate the cost of work by calculating the cost from the labor, material, equipment, and vehicles used on a job. It is also possible to roll up the cost of maintenance for work types and asset types. This functionality is also useful in developing the utility's budgets.

5.3 Compliance Monitoring and Reporting

All utilities must comply with federal water quality standards and submit reports to the Environmental Protection Agency. The data associated with compliance reporting may be stored in a number of different IT systems and/or on logs kept by staff. Operational data systems, including sensors and laboratory information management systems (LIMS), are the primary source of data, and some LIMS have compliance management capabilities.

There are also a variety of software products on the market to support this necessary business activity and provide capabilities to share this information with the public. This is especially important when an event occurs that triggers a water quality concern and timely communications can affect public health. There is typically a designated staff person responsible for regulatory compliance and they are the primary user of the system, but the data may be generated by operational staff or systems. Data typically include the regulatory standards for water quality, and the observed water quality values over time and on average for the system.

5.4 Project and Program Management

Project management is a key function within a utility. It requires coordination across different stakeholders and generates many documents and files that must be accessed by team members both in and outside of the utility. The core of project management includes scope, schedule, and budget, so the systems must support all of these things. A robust project management system will facilitate communications and information sharing to keep projects on schedule and on budget.

Reports and views are common and crucial requirements as the system is used. Some examples of reports are cash flow; various types of estimates; budgets (current, baseline, or projected) and schedules (Gantt charts and critical path reports); resource allocations; financial commitments; cost to date; spending rates; funding; design and construction status to date; various project status indicators; change order logs; change order reasons; planned versus actual costs; and issues and action items reports. Accuracy and availability of these reports is important for timely decision making and successful project management.

The term *project management system* covers a range of software, including scheduling, cost control and budget management, communication and collaboration software, quality management, project documentation or administration software, and resource allocation. Project management systems can be delivered through a variety of system architectures and can range in use and complexity from single-user systems running on a single desktop computer to collaborative, multiuser systems that integrate project planning, project control, and management functionality. Project management systems typically come with a built-in reporting system and dashboards for information visualization. It is important to state the differences between a project management system and an enterprise resource management system. An ERP system is implemented and used to manage resources, activities, and information for an organization; whereas project management systems are used for managing information, resources, and activities for a particular project or sets of projects/programs. Project management systems are used for planning and capturing of project information in real time, whereas this information becomes part of an ERP system after the project is realized.

End users of a project management system vary depending on the module used and end users' job responsibilities. Users of a project management system include project controls staff (schedulers, cost engineers, estimators, and contract administrators); field staff (inspectors, superintendents, and supervisors); project management staff (project engineers and project management); and senior management. The level of detail and information presented change based on the end user's role in a project. For example, a scheduler might want to view tasks, their dependencies, and resources required by each task in a particular project, whereas a project controls manager might want to view schedules of various projects.

The types of data that are associated with different project management system modules are shown in Table 2.3. The data presented in Table 2.3 demonstrate the overlaps in data used for different uses within a project management system. The modules of functionality do not operate in isolation, and integration between data sets for multiple uses is an important component of such a system.

Although project managers implicitly recognize the importance of time and cost information integration for successful project management, it is rare to find effective project control systems that include both elements. Typically, project costs and schedules are recorded and reported by separate application programs. Project managers must then perform the tedious task of relating the two sets of information. The difficulty of integrating schedule and cost information stems primarily from the level of detail required for effective integration. Typically, a single project activity will involve numerous cost account categories. Similarly, numerous activities might involve expenses associated with particular cost accounts. To integrate cost and schedule information, a common work breakdown structure and specific activities and specific cost accounts must be the basis of analysis.

Another area of integration is between project management systems and ECM or enterprise resource management. Archiving project information when a project is completed is typically a daunting task if done manually rather than automatically. In addition, databases recording the "as-built" geometry and specifications of a facility as well as the subsequent history can be particularly useful during the use and maintenance life-cycle phase of the facility. As changes or repairs are needed, plans for the facility can be accessed from the database.

Project management system vendors offer many types of hosting options, such as the following: use of an ASP, also called *on-demand software* or *SaaS*, where the software vendor provides computer-based services to customers

TABLE 2.3 Types of Data Associated With Different Project Management System Modules

Module	Data
Scheduling	Work breakdown structure and tasks Tasks with durations and dependencies Constraints Resources Cost for each activity Milestones
Cost estimating	Work breakdown structure Material quantities and prices Equipment Labor Historical data/cost-estimating reference data Durations
Cost control	Work breakdown structure Budgets Cost accounts Payables Receivables Incurred costs
Contract management	Contract documents Change requests Change orders Amendments Invoices Payment applications
Document control	Project documents, including • Permits • Meeting agenda • Meeting minutes • Correspondence • Disputes • Submittals • Specifications • Memorandums • Regulatory documents

over network and host servers; self-hosting, where the customer hosts servers in-house; or third-party hosting, where customers outsource hosting. Deciding between these hosting options depends on security, retention, storage, bandwidth, and speed needs, as well as the availability of an organization's technical support group. If an ASP model or third-party hosting option is chosen, then, at a minimum, a license should be maintained to access project information or archiving options should be discussed with the service provider. In essence, how to maintain data and access data when a project is completed are important topics to be clarified before investing in any project management system options.

5.5 Construction Management

Construction management has evolved significantly in the past decade in terms of the use of IT to support planning, design, construction, and commissioning of assets. Although computer-aided drafting and design software has been in use for decades, its evolution to BIM, 3-D modeling, and VDC has provided great opportunities. Utility staff and their consultants have become more proficient, and software vendors are providing greater capabilities that allow a full life-cycle consideration of an asset. When a new facility or system component is designed, or when a facility or system component is overhauled, it is possible to complete the design in a manner that facilitates the construction or rehabilitation of the asset, and to capture the as-built data for future use in the asset management system.

Virtual design and construction has emerged as a best practice to improve the design process by providing highly detailed multidimensional models so the utility staff can better envision the physical system before it is constructed. This results in fewer change orders and faster project delivery time. There are emerging technologies that complement this including virtual/mixed and augmented reality so people can immerse themselves in the new physical area virtually to better understand the design.

The data in these systems includes the identifier of and key attribute for each physical component of the design, which increases in volume and detail as the design progresses. The users of these systems include utility project staff and the staff responsible for the future operations and maintenance of the assets.

A notable data collection process for utilities that are expanding or rehabilitating their systems is the use of LiDAR technology to use light from lasers to locate existing assets. This can be used to create complete digital models of current as-built conditions with greater accuracy and completeness than may be available within the utility's records systems. This not only can benefit construction rehabilitation efforts but can also be valuable for asset management.

6.0 OPERATIONS

Operational systems are those systems that interface directly with the water and wastewater systems to monitor and manage operating conditions and enable adjustments to respond to changes in conditions. There are three general categories of operation systems that are important to understand because they are part of a utility's overall IT infrastructure needs, and because data from these systems are being combined with data from IT systems to support advanced analysis. As computing capabilities continue to grow exponentially, analyzing extremely large data sets is becoming more accessible for utilities to better understand and predict system performance and make well-informed capital investment decisions.

6.1 Collecting Operational Data

There are a number of different parameters that utility operators monitor to ensure the facility is operating as expected. This includes flowrates, temperature, pH, and a wide variety of water quality measures. There are four elements to the process that each have supporting devices and integrations between them:

- measurement (by sensors or online analyzers)
- recording output signals (logger unit)
- uploading/accessing recorded data (telemetry)
- analysis of recorded data (data acquisition software)

Operational technology includes the devices, networks, and software to measure, record, upload, and analyze data. These systems are explored in depth in WEF's MOP 21 and will not be discussed further here, but it is useful to understand the role of this technology because the data that are collected are used directly or indirectly with IT systems to manage the utility and serve customers. Chapter 7 will describe the context for and relationship between operational technology and IT in designing the overall system architecture for a utility.

6.2 Water Quality Testing

Water and wastewater utilities have water quality standards to which they must abide and report to their regulatory authority. When water quality does not meet standards, this must be reported to regulators and customers. Water quality testing is done throughout the system, with more frequent measurements done within the facility, at outfalls, and in the distribution system. Testing can be done using technology, or manually through controlled sample management procedures. Some samples may be analyzed at point of capture, in a utility operated laboratory, or they may be sent to an outside laboratory depending on the capabilities of the utility. All samples are managed according to strict chain of custody rules to ensure the efficacy of the samples and associated results. The data from these processes are necessary for regulatory reporting, customer communications, and evaluating system operations.

A laboratory information management system (LIMS) is a type of software that supports the recording, storing, analyzing, and reporting of the results of laboratory analyses. It provides information about analytical samples received and tested within the laboratory operation. A LIMS can provide information regarding analytical results, status of testing in progress, sample collection data, workload, summary reports and trend analyses for sample analytical results and business operations, and quality control information. Data entry to the LIMS can be manual or automated through integration of laboratory instrumentation. A LIMS can be configured to import data from standardized external sources such as spreadsheets. The data collected include environmental conditions associated with the time of the sampling, location of sample, and relevant water quality results depending on the tests run. This data can be insightful when building models to understand and optimize operations. Some LIMS include capabilities for operational analysis including trend charts, correlation analyses, and alarms or notifications when there are outliers or changes in trends from the norm.

The primary end users of the LIMS for data entry are the utility's laboratory technicians or facility operations staff responsible for sampling. They are designed to replace the technician's logbook. In practice, staff may enter the data directly to the system or transfer data from logbooks to the LIMS. Additional stakeholders are users of outputs from the LIMS, including facility operations staff responsible for adjusting operating parameters to address water quality, facility management as part of benchmarking and monitoring performance, and those staff responsible for regulatory reporting.

6.3 Operational Control

System operations involve a complex and dynamic network of physical, digital, and human processes. As external conditions such as weather, influent quality and quantity, and regulatory requirements change, system operators in the facility and in the field must respond quickly to protect the health of their communities.

Supervisory control and data acquisition (SCADA) systems and their human–machine interface (HMI) are the most prevalent operational technology systems used in water and wastewater facilities, pumping stations, and collection and distribution systems. Through frequent data collection via remote sensors, operations staff can have a system-wide view in real time through the HMI so warning signs can be observed more quickly. Some components of the system may also be controlled through the system to enable faster and more efficient reaction time. The primary users of these systems are operations staff, and utility directors may have view access so they can be aware when time critical events occur. The data in these systems vary by utility. Some may have a minimal set of monitoring and control points, whereas others may have more complete digital coverage of their physical system.

Process control is not only about running equipment such as pumps, compressors, and valves. It is also about consistently meeting the requirements of the operation while minimizing operating costs. This means that the facility system has to be understood from a dynamic point of view. It is crucial that operators are part of both the design process and the control system definition to capture their understanding of the system, and to build the trust necessary to facilitate effective use of the system when it is deployed.

The primary inputs to a control system are measurements from online sensors. Measurements of adequate process variables have to be provided with adequate sampling rates. Monitoring is an important part of the system and can provide critical information about the facility as well as present early warning signs of disturbances and process changes. The actuators, such as pumps, valves, and compressors, have to be designed so that the facility is truly controllable. Indeed, too many control systems have failed because of inadequate actuators. Only the team at the facility can provide true improvements to the operation.

The Water Environment Federation's MOP 21 discusses these systems in detail, but we will discuss some basics here because the data and business processes associated with operational control are an important component of a utility's IT program. The topic of cybersecurity related to these systems is covered in Chapter 8.

7.0 PLANNING AND DECISION SUPPORT

The use of IT and the data from a utility's IT software is evolving rapidly. At the time of this writing, there is much effort to explore the role of artificial intelligence and digital twins to support utility operations and management. As outside pressures from climate change and failing infrastructure continue to strain utility resources, it is becoming more and more important to apply analytical modeling tools to support utility planning and decision making.

The foundations for these efforts have been in use for many decades, and those efforts will be discussed in this section.

7.1 Planning

Utility planning efforts typically occur as a lead-in to the annual budgeting process. This is a time to review operations and maintenance costs and performance and define longer term capital project needs. Asset management systems and financial systems are the primary inputs to these efforts. A supporting IT system for this is the GIS. A utility GIS is an important tool for managing information about the utility's assets and customers. Indeed, wherever location is important or maps are necessary to conduct business for the utility, there is an important role for GIS. The Water Environment Federation's Manual of Practice 26, *GIS Implementation for Water and Wastewater Treatment Facilities,* describes the nine fundamental purposes of a GIS that provide value to users as follows: "mapping and databases, facilities management, facilities atlases, management decision making, facilities planning, federal regulation compliance, business process reengineering, public perception, and e-commerce" (WEF, 2005). Figure 2.4 presents an example of the data tables related to the spatial map interface.

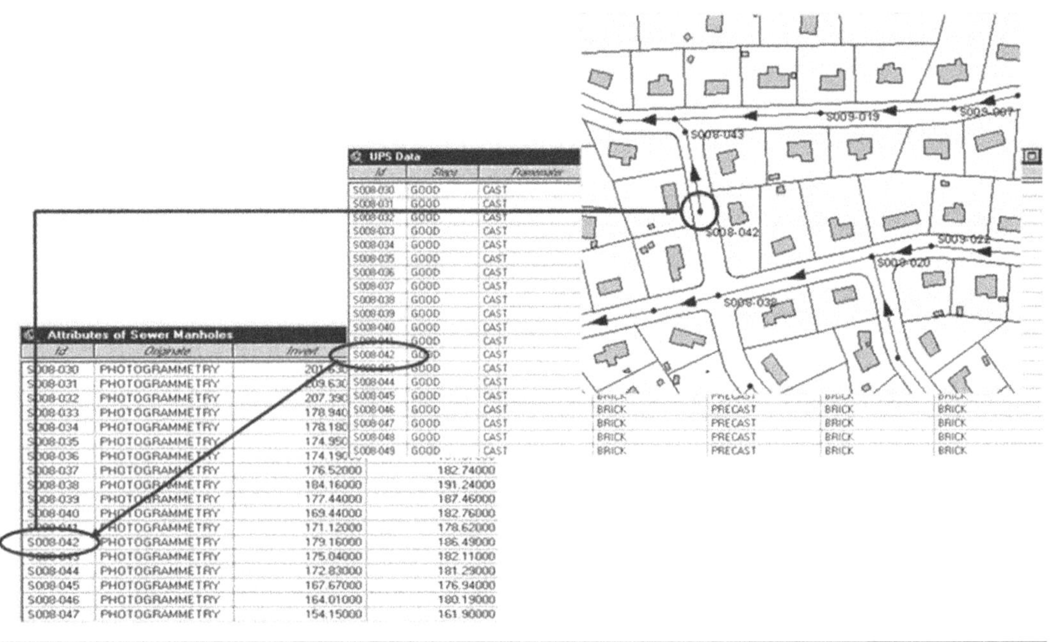

FIGURE 2.4 Spatial Data Relationships

Because of the potential variety and extent of GIS use, there can be end users in many business areas of the utility as well as with stakeholders outside of the utility. Typical end users interact with the data to perform queries and analysis and to create map-based outputs. There is a smaller subset of end users responsible for adding and editing GIS data. To ensure data quality and integrity, it is important to have trained GIS professionals responsible for maintaining system data. Below is a summary of some of the general uses of a GIS for a utility.

Geographic information system professionals are the staff responsible for adding new data to the GIS. The creation and editing of data layers and features in a data layer requires training and experience in the placement of data, the maintenance of metadata, and core geographic principles. The Urban and Regional Information Systems Association (Des Plaines, Illinois) (www.urisa.org) has programs to train GIS professionals; the GIS Certification Institute (Des Plaines, Illinois) (www.gisci.org) also provides certification for GIS professionals.

Field use of a GIS can be beneficial for staff working in the field conducting inspections, maintaining assets, and sampling. It is becoming more common for field staff to carry laptops or handheld devices while conducting their work, and having internet access or GIS files on those devices can help them to more accurately and efficiently locate spatial information necessary to complete their tasks.

At the management level, a GIS can become a key tool to support utility decision making and planning if the GIS is well integrated with other data sources. This can include capital improvement planning and project schedule, public outreach, permitting, and land acquisition. In addition, a GIS is a useful tool when working with other utilities, such as phone, energy, and cable companies, to coordinate road cuts and other project work. Figure 2.5 presents a view of a map with multiple layers, or sets, of information that can be viewed simultaneously to support interdepartmental collaboration.

Finally, a GIS can be a powerful tool for communicating with the public. Many governmental entities such as states, cities/towns, and counties now manage extensive GIS data sets and publish them on the internet for the public to query. There are also private companies, such as Google (Mountain View, California), that provide extensive GIS content through the internet. Many utilities do not publish their data on the internet, often because of security concerns. However, some layers of information are often used to communicate with the public about large upcoming projects that require public support or that would have an effect on residents. This may be done through GIS capabilities online, such as queries, or through the publication of static maps created with a GIS. Additional details about GIS can be found in MOP 26.

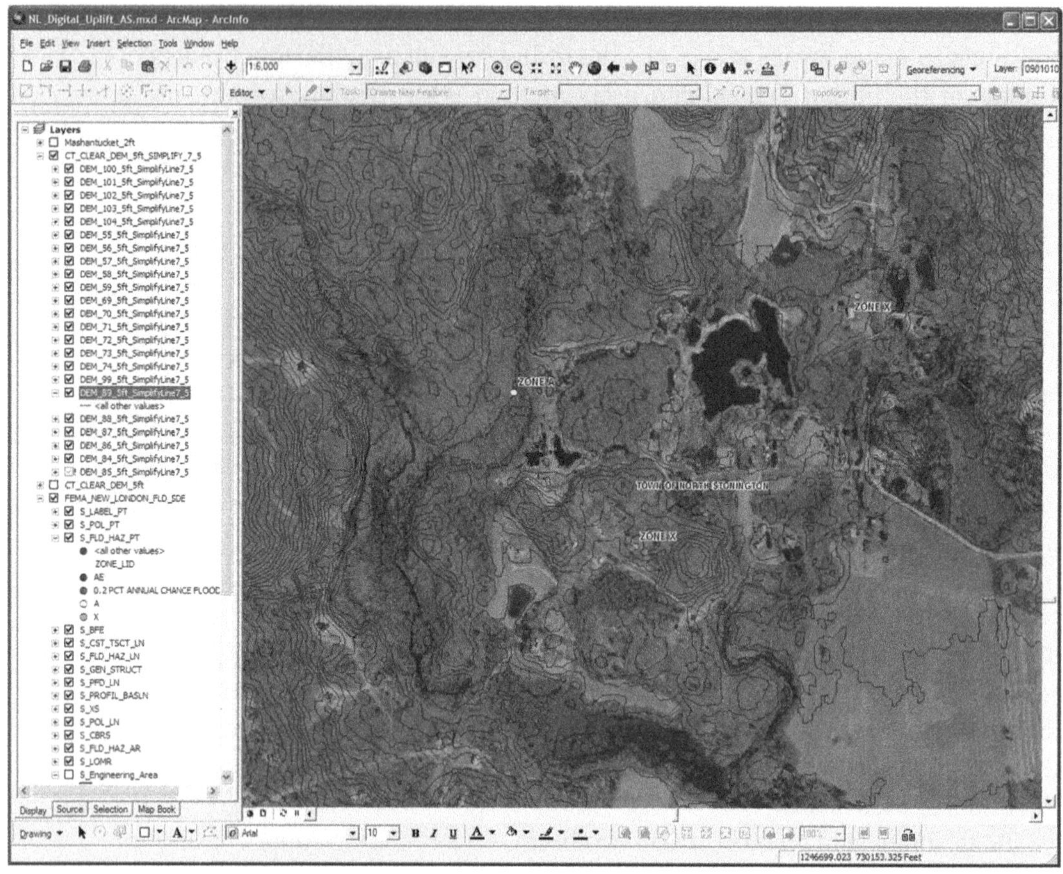

FIGURE 2.5 Multidimensional Spatial Data

7.2 Modeling

Models provide idealized representations of an actual physical system. Models are typically used in a general sense to evaluate the operation of a system or process under actual (typically for troubleshooting) or theoretical (typically for planning) conditions. Well-developed and calibrated models allow a utility to simulate "what if" processes for decision support purposes before attempting changes in the physical system. In summary, mathematical models allow utility stakeholders to both understand how a system is operating now and how it might operate under a different set of conditions. Classic applications include planning for future development (how will my system react to the addition of future customers?), out-of-service scenarios (how can I operate the system with this component removed from service for repairs?), and optimization scenarios (what can I change to better meet a set of operational goals?). Commercial off the shelf models are available for all types of system evaluations. The most commonly applied are water distribution models, wastewater and stormwater conveyance (or hydrology/hydraulics) models, and process optimization models (for evaluating treatment processes). Each of these is covered separately in the following sections.

7.2.1 Water Distribution Models

Water distribution models represent a specific genre of hydraulic model applied to systems operating under pressure. These may include raw water supply, finished water transmission and distribution, manifold force main wastewater systems, and water reuse/delivery systems.

End users of water distribution models are typically limited to engineers, planners, and operations staff. Most models are somewhat complex and, therefore, demand a certain familiarity and repetitive practice to be used effectively. Hence, it is important that staff using hydraulic models be provided both sufficient training and time to accomplish hands-on activities.

Whereas water distribution models can perform numerous functions requiring many types of data, most analyses can be performed by assembling the following six types of data: (a) physical data (diameter, length, roughness coefficient) on the pipes in the system; (b) consumption data that define the demands on the system, that is, the average volume of water to be delivered to each customer; (c) pattern data that define the variability in water demand over the course of a day; (d) elevation data for each feature that enable model calculations of hydraulic grade line to be expressed as a pressure; (e) data for critical features in the system that supply and/or moderate hydraulic grade line, including pumps, tanks, wells, and specialty valves; and (f) operating rules that define how tank levels, pressures, and/or other system parameters will govern the operation of model features, particularly critical features such as pumps.

The most basic water model output includes flow, velocity, and head loss in each pipe and pressure at each junction. These can be for a single point in time (for a steady-state model) or over a time series (for an extended-period model). Models will also provide other results such as system-wide demand, theoretical fire flow availability at any or all node locations, water age, water quality for a conservative or reactive substance, pump energy costs, and so forth. Model output can be used to support utility decisions ranging from how large a replacement pipe should be to the best way to serve an area of low pressure, to how to operate a series of pumps most efficiently to minimize electrical costs.

Because of the variety of data inputs required to enable a functioning model, water distribution models require data from, and can be integrated with, several other systems commonly found in utilities. Table 2.4 illustrates key integration areas.

TABLE 2.4 Key Integration Areas for Water Distribution Models

System	Water Model Integration Areas	Purpose of Integration
Automated meter reading	Customer demand patterns	Develop individual or representative demand patterns for large users or user classes, respectively
Customer information or billing system	Customer locations and demands	Develop necessary essential model for solving system hydraulics
BIM	As-built features/assets	Support development of physical network if GIS data are not available or inaccurate
GIS	As-built features/assets	Develop physical network of pipes and junctions
SCADA	Operational aspects	Validate system operations Support development of diurnal demand patterns

Software maintenance is typically straightforward, and an annual maintenance fee will enable new versions to be identified automatically and easily downloaded and installed. Data maintenance presents unique challenges for hydraulic models, primarily in the area of GIS integration. This is because the model is not operated directly against enterprise GIS data and also because models tend to be used to evaluate many what-if scenarios that involve the addition of assets that are not present in GIS data. As a result, two-way synchronization schemes are not viable without significant checking and/or customization to avoid unintended consequences (such as proposed model features being deleted because they do not appear in the GIS).

7.2.2 Hydrology and Hydraulics Models

Hydrology and hydraulics (H&H) models represent a specific genre of hydraulic model applied to systems that are, for the most part, intended to flow under gravity conditions rather than under pressure. These may include open channel/riverine systems, sanitary wastewater collection systems, stormwater collection systems, and/or combined sewer systems. In many respects, H&H models are similar to water distribution models. The following sections, therefore, concentrate on some of the major differences of H&H models.

Like water distribution models, end users of H&H models are typically limited to engineers, planners, and operations staff. As opposed to water distribution models, many H&H models, particularly fully dynamic models, tend to be more subject to instabilities that can require special expertise to understand and correct.

Hydrology and hydraulics models generally require the same types of physical data as water distribution models. One major difference is that H&H models typically place less emphasis on individual customer demands and patterns (an exception is the unusual case in which even dry weather flows tax the gravity system) and much more emphasis on rainfall and runoff and/or the effects of rainfall-induced infiltration and inflow. As a result, riverine models require additional data to be collected as well as input related to rainfall and land cover characteristics, especially imperviousness and soil types/characteristics. Closed-system models require information on the levels of rain-dependent infiltration and inflow expected to enter the system. Other differences include the need for cross-section data in riverine systems, the inclusion of support for a variety of pipe shapes (especially in older systems), and the introduction of additional specialty features such as weir and orifice elements.

Hydrology and hydraulics model output primarily includes time series data on flows, velocities, and depth of flow across the system. Like water models, some H&H models can optionally support water quality evaluations such as dissolved oxygen analysis in streams. Model output can be used to support utility decisions ranging from how large a replacement pipe should be to the best way to serve an area of local flooding, to how to operate a series of control structures to minimize combined sewer overflows.

Like water distribution models, H&H models require data from, and can be integrated with, several other systems commonly found in utilities. Table 2.5 illustrates the key integration areas of H&H models. Software and data maintenance requirements are similar to those for water distribution models.

7.2.3 Process Models

Process models are commonly used in the water and wastewater industry. A *process model,* in this context, is defined as a mathematical formulation of chemical, biological, or physical processes that occur in water and wastewater process tanks. These models simulate processes that take place in water and wastewater process tanks so that the behavior of full-scale processes can be predicted. The models are implemented via computer software for use in planning, design, and operation of water and wastewater conveyance, storage, and treatment facilities. Process models are used in the planning and design of water and wastewater treatment facilities to size process tanks, to predict removal of pollutants and changes in chemical parameters during

TABLE 2.5 Key Integration Areas for H&H Models

System	Water Model Integration Areas	Purpose of Integration
Automated meter reading	Customer demand patterns	Develop individual or representative demand patterns for large users or user classes, respectively
Customer information or billing system	Customer locations and demands	Establish base flows in wastewater system models
BIM	As-built features/assets	Support development of physical network if GIS data are not available or inaccurate
GIS	As-built features/assets	Develop physical network of pipes and junctions Develop areal statistics for model support such as impervious area and soils characteristics
SCADA	Operational aspects	Validate system operations Collect rainfall data

treatment such as alkalinity and pH, and to optimize tank geometry. They can be used for operator training, as decision-making guides, or to generate values for control of real-time systems.

Process models may be characterized in several ways. One way to characterize them is to ask whether they attempt to predict what goes on within a process tank by means of a mechanistic method; another way to characterize process models is to ask whether they incorporate probabilistic methods based on prior behavior of full-scale processes. Models like the simulation of single sludge processes model (Clemson University, 1987) of the activated sludge process fall into the former category. Models like the autoregressive integrated moving average (ARIMA) model (Box & Jenkins, 1970) and the Monte Carlo–type model (Metropolis & Ulam, 1949) fall into the latter category.

Another way to characterize process models is based on the degree of knowledge about the underlying flow field that they incorporate. The following three classes of models are identified in this way:

- *Black-box models.* These types of models assume no knowledge about the spatial variation of process parameters within a process tank.

Examples of this type of model are residence time distribution theory and the ARIMA model. These models rely entirely on empirical evidence from instances of similar full-scale process configurations for calibration. Their predictive value results from the fact that outputs are typically related to inputs. However, they are disadvantageous compared to models based on the equations of fluid mechanics and biological and chemical kinetics in that they cannot be applied in instances where data from similar applications cannot be obtained. Their predictions may depend on hydraulic residence time in the tank but make no allowance for differing features of tank geometry.

- *Gray-box models.* This intermediate type of model uses a simplified model of tank flow within the reactor to predict chemical and biological changes within the tank. Examples of this type of model are the dispersion model and the tanks-in-series model. The dispersion model permits flow in one dimension only. It is sometimes called the *dispersed plug-flow model.* Turbulent spreading or dispersion is modeled in either axial or radial directions in pipes and fluidized beds. When dispersion in the axial or longitudinal direction is modeled, it is called the *axial dispersion model.* The tanks-in-series model assumes that transport in a process tank can be modeled by replacement of the real tank by an "equivalent" number of completely mixed single tanks operated in series. This type of model has become widely used for modeling of activated sludge wastewater treatment processes.
- *Glass-box models.* This type of model bases its prediction of tank process behavior on calculation of the multidimensional velocity field within the tank. These models place solids transport or chemical reaction models within a flow field, which has been predicted using mathematical equations of fluid mechanics in two or three dimensions. By predicting the variation in velocity and pollutant concentration within the tank using the equations of computational fluid dynamics (CFD), these types of models can more accurately predict process behavior than the previous two types of models. Models based on CFD have been used for planning and design of sedimentation and disinfection tanks and for analysis of flow problems in treatment facility hydraulic elements. The disadvantage of this type of model is that it can be compute intensive and requires high-level knowledge of the underlying fluid mechanical theory for effective use.

In general, black-box models are stochastic and gray- and glass-box models are mechanistic, in the sense defined previously.

Process models are typically used today by engineers in planning and design of water and wastewater process facilities or for the diagnosis of problem behaviors in full-scale facilities. However, process models can also be effectively used for operator training and for day-to-day decision making in water and wastewater treatment facilities. During the 1990s, the Commonwealth of Massachusetts used wastewater treatment simulation software based on black-box models for operator training. Many treatment facilities have used dynamic models of facility hydraulics and tanks-in-series models of facility biological processes for day-to-day decision making.

Process models use inputs from water or water resource recovery facility operation or from projections of future facility operation to simulate future behavior. These data include

- facility flow;
- site characteristics such as elevation, air temperature, and wind speed;
- process fluid characteristics such as temperature and viscosity;
- process chemical characteristics such as pH and alkalinity;
- pollutant concentrations such as turbidity, suspended solids, biological and chemical oxygen demand, and ammonia; and
- tank and conveyance element dimensions.

Process model outputs typically include predictions of effluent characteristics from process tanks such as

- facility flow;
- process chemical characteristics such as pH and alkalinity; and
- process pollutant concentrations such as turbidity, suspended solids, biological and chemical oxygen demand, and ammonia.

Process models are also used to optimize the geometry of process tanks such as sedimentation tanks, disinfection tanks, and hydraulic components. This is done by simulating the behavior of the tanks under different geometric arrangements to predict which geometric features will be most effective during treatment. In this case, the outputs from the process models are comparative profiles of velocity, solids, and/or reactant concentrations within the tank.

Process models can be used to "fill in" missing data for process control of systems such as activated sludge aeration or sludge wasting. They can be included in the dashboard of facility operations personnel to aid in decision making.

Process models are available from a wide variety of sources, including academic, proprietary, commercial, and public domain. Examples of black-box types of models, which have been used in the water and wastewater industry, include neural network models, ARIMA models, and Monte Carlo simulations. Neural network models have been used to forecast inflow into storage tunnels; ARIMA models have been used for evaluation of clarifier performance and for linearization of flow networks. Monte Carlo models have been used for the evaluation of process data for the selection of appropriate peaking factors for wastewater treatment planning. These have all been implemented as proprietary software development projects by individuals or consulting engineering companies.

The software typically includes convenient aids to geometric definition of flow grids, a variety of models for turbulence, automatic generation of contour and other types of plots to enhance visualization of results, and the capability for customization by the addition of a user-defined function. For proprietary models, maintenance of process modeling software is done by in-house programmers and, for commercial products, via updates and licensing agreements with vendors.

This area is changing rapidly, and new modeling tools are emerging. It should be noted that developing a model that effectively represents system conditions requires investment and takes time. The choice of modeling tool, the rules/relationships, reference data, and extent should be well thought out to ensure that the model can be used over the future. Utilities often engage engineering consultants to support model development and maintenance, but it is important that the utility staff responsible for applying the results have a solid understanding of the assumptions, limitations, and strengths of the model to ensure the decisions are properly informed.

7.3 Deciding

Utility staff across the organization make time-critical decisions on a daily basis. Some of these decisions can be made with information from a single IT system, but often data from multiple systems must be brought together to provide the necessary insight and context. An information portal is an IT interface that integrates data from multiple sources to provide end users with a unified perspective of the information. A web-based interface presents summaries of data to meet specific, customized business needs that cannot be met with a report from a single information system. The screens that users interact with and view are often referred to as "dashboards." Commercial enterprise systems that include multiple modules of data to support a variety of business processes may come with portal functionality that is configurable to the organization's needs. Alternatively, portals may be customized to integrate disparate information silos.

Portals are an effective mechanism to provide outputs of information to end users. The inputs to portals are the databases of organizations that are part of other information systems; data are typically inputted directly through technology integrations rather than entered by end users. The content of the inputs and outputs is configurable or customizable based on the information, metrics, and decision-making needs of the utility.

Dashboards can be developed for many different end users with different responsibilities and perspectives to provide information that is operational, tactical, and strategic. In all these instances, the idea is that end users are decision makers that need access to information and metrics that enable them to move the organization closer to an agreed-upon set of goals in a timely manner. To do this, it is necessary to have access to data that is high quality, timely, and in line with the organization's metrics and goals.

Figure 2.6 presents a dashboard that provides facility operators with a bird's-eye view of the entire distribution system. It includes information about alarms that are triggered based on flow and pressure trends. This type of configurable alarm dashboard can help operators make proactive decisions based on real-time data. One of the greatest benefits of a dashboard is the ability to target information for specific decision makers. For example, the dashboard designed for a utility's operational purposes can support end users who are front-line workers and supervisors responsible for monitoring and optimizing operating processes. It can include diagnostic metrics about flowrates, energy consumption, and water quality that are frequently updated to provide the information needed to best run the facilities and associated system components.

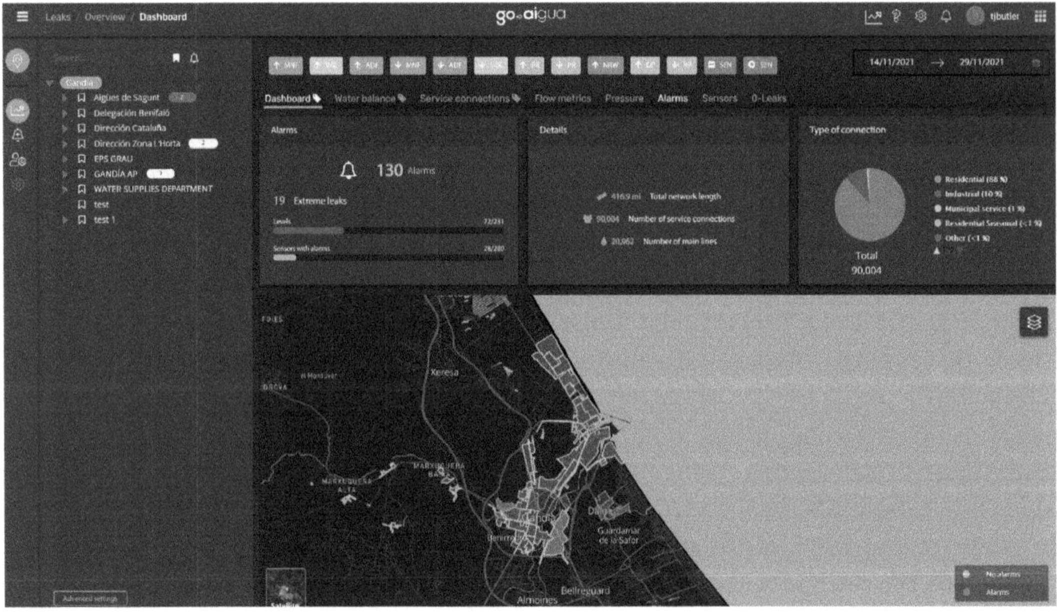

FIGURE 2.6 Dashboard of Collection System (Credit GoAigua, Inc.)

A dashboard designed for tactical purposes typically targets different supervisors and managers who require self-service access to information with direct navigation to monitor progress of metrics against baselines and goals. It can also be possible to enable end users to "drill down" their analyses by targeting a specific metric result and obtaining the details of the calculation behind it. The calculations may reside in the source system or the portal. Data are typically updated less frequently than in an operational system and, depending on business needs, may be daily or weekly.

Strategic dashboards are developed for use by senior management responsible for measuring high-level objectives and making complex decisions based on qualitative and quantitative information. The information presented typically focuses on longer term trends, and data may be compiled monthly.

Because portals are, by definition, mechanisms to integrate data sources, the type and number of systems that may be integrated is driven by business needs. Because of the complexity of data and system integration, for a portal development project to be successful it is important that sufficient attention be given to the ultimate information and decision-making needs of the organization. Outputs and functionality must be useful and ambitions for integration must be realistic. The integration needs between data sets must be well thought out in addition to the methods and business rules for compilation, calculation, and presentation.

System maintenance considerations for portals depend, to a great extent, on the systems that feed the portal and the technologies applied to develop the portal. For a utility that has established technology standards with common database software and platforms, portal development can be done in a manner that minimizes the maintenance burden. Most maintenance requirements will develop from new user requirements for outputs, changes in business rules, and replacement of the core systems that feed the portal. As with all utility systems, there should be a business owner or set of owners who participate in a regular process to evaluate the portal's ability to meet business needs, and a budget should be set for periodic maintenance to keep the portal current and to maximize its effectiveness for the organization.

8.0 REFERENCES

American Water Works Association. (2012). *M6: Water meters—Selecting, installation, testing & maintenance* (5th ed.).

American Water Works Association Research Foundation. (2006). *Asset management planning and reporting options for water utilities.*

American Water Works Association Research Foundation. (2008). *Asset management research needs roadmap.*

Box, G., & Jenkins, G. (1970). *Time series analysis: Forecasting and control.* Holden-Day.

Clemson University. (1987). *Simulation of single-sludge processes for carbon oxidation, nitrification, & denitrification.* Retrieved August 2009 from http://www.clemson.edu/ces/departments/eees/outreach/sssp.html

Kavanagh, S. C., & Miranda, R. A. (2005). *Technologies for government transformation: ERP systems and beyond.* Government Finance Officers Association.

Metropolis, N., & Ulam, S. (1949). The Monte Carlo method. *Journal of the American Statistical Association, 44*, 335–341.

Twigg, R., Payne, J., & Tzaban, Y. (2007). *IBM information management software, ECM taxonomy management.* Retrieved March 2010 from ftp://ftp.software.ibm.com/software/data/cmgr/pdf/omnifind-DE-taxonomy.pdf

Value of Water Campaign and American Society of Civil Engineers. (2020). *The economic benefits of investing in water infrastructure.* http://www.uswateralliance.org/sites/uswateralliance.org/files/publications/The%20Economic%20Benefits%20of%20Investing%20in%20Water%20Infrastructure_final.pdf

Vitasovic, C., Olsson, G., & Haskins, S. (2021). *Water intrapreneurs for successful enterprises (WISE): A vision for water utilities.* Water Environment Federation, Leaders Innovation Forum for Technology, & Water Research Foundation.

Vitasovic, C., Olsson, G., Ingildsen, P., & Haskins, S. (2022). *Water intrapreneurs for successful enterprises (WISE): A vision for water utilities.* Water Environment Federation & International Water Association.

Water Environment Federation. (2005). *GIS implementation for water and wastewater treatment facilities* (Manual of Practice No. 26).

Water Environment Federation. (2013). *Automation of water resource recovery facilities* (4th ed., Manual of Practice No. 21).

9.0 SUGGESTED READINGS

American Water Management Association, National Association of Clean Water Agencies, & Water Environment Federation. (2007). *Implementing asset management: A practical guide.*

American Water Works Association Research Foundation. (2005). *Effective practices to select, acquire and implement a utility CIS.*

Baumert, J., & Bloodgood, L. (2004). *Private sector participation in the water and wastewater services industry's office of industries* [Working Paper No. ID-08]. U.S. International Trade Commission.

Boyer, S. A. (2004). *SCADA: Supervisory control and data acquisition* (3rd ed.). ISA Press.

Godin, F., Varner, J., Peters, S., Brueck, T., & Brink, P. (2015). *Adapting leading practices and associated tools*. IWA Press.

Griborio, A. (2004). *Secondary clarifier modeling: A multi-process approach* [PhD thesis, University of New Orleans].

Griggs, N. S. (1996). *Water resources management: Principles, regulations, and cases*. McGraw-Hill.

Henze, M., van Loosdrecht, M., Ekama, G., & Brdjanovic, D. (Eds.) (2008). *Biological wastewater treatment: Principles, modeling and design (UNESCO)*. IWA Publishing.

Hong, Y., Zhang, Y., & Khan, S. L. (2016). Hydrologic remote sensing: Capacity building for sustainability and resilience. CRC Press.

International Organization for Standardization. (2003, October). *Water quality: On-line sensors/analysing equipment for water—specifications and performance tests* (1st ed.). ISO 15839:2003. International Organization for Standardization.

International Society of Automation. (2003). *Automation, systems and instrumentation dictionary* (4th ed.). ISA Press.

International Water Association Task Group on Mathematical Modeling for Design and Operation of Biological Wastewater Treatment. (2000). *Activated sludge models*. IWA Publishing.

McCorquodale, J., La Motta, E., Griborio, A., Homes, D., & Georgiou, I. (2004). *Development of software for modeling activated sludge clarifier systems: A technology transfer report*. University of New Orleans Department of Civil and Environmental Engineering.

Olsson, G., & Newell, B. (1999). *Wastewater treatment systems: Modeling, diagnosis and control*. IWA Publishing.

Olsson, G., Nielsen, M. K., Yuan, Z., Lynggaard-Jensen, A., & Steyer, J. P. (2005). *Instrumentation, control and automation in wastewater treatment systems* (Scientific and Technical Report No. 15). IWA Publishing.

Ramon, G., & Stern, C. (2004, March 14–17). *Lessons learned in implementing mobile computing technology to improve utility field operations* [Paper presentation]. *Proceedings of the 2004 Water Environment Federation/American Water Works Association Joint Management Conference*, Phoenix, AZ, United States.

Ramos, L., Orlov, L. M., & Teubner, C. (2004). *Making portals work*. Forrester Research, Inc.

Rosen, C., Ingildsen, P., Guildal, T., Nielsen, M. K., Munk-Nielsen, T., Jacobsen, B. N., & Thomsen, H. (2005). Introducing biological phosphorus removal in an alternating plant by means of control: A full scale study. *Water Science & Technology*, *53*(4–5), 133–141.

Segal, G. F. (2003). *Georgia public policy foundation, issue analysis: The Atlanta water privatization: What can we learn?* Retrieved August 2009 from http://www.gppf.org/article.asp?RT=20&p=pub/Water/atlanta_water.htm

U.S. Environmental Protection Agency. (2017, January). *Effective utility management: A primer for water and wastewater utilities.* watereum.org

U.S Environmental Protection Agency. (2021). *Resources to incorporate advanced metering infrastructure into surveillance and response systems.* https://www.epa.gov/waterqualitysurveillance/resources-incorporate-advanced-metering-infrastructure-surveillance-and

Vitasovic, Z., Zhou, S. P., & McCorquodale, J. A. (1997). Secondary clarifier analysis using data from the CRTC protocol. *Water Environment Research*, *69*(5), 999.

Water Environment Federation. (2013). *Wastewater treatment process modeling* (2nd ed., Manual of Practice No. 31).

Water Environment Federation. (2020). *Stormwater, watershed, and receiving water quality modeling* (WEF Special Publication).

3

The Importance of Data

1.0 SUMMARY OF KEY THINGS TO KNOW

The intent of this chapter is to highlight the value of the data associated with information technology (IT) systems. Considering data as an asset is relatively new for utilities as our economy has transformed. Tools to manage and analyze data have matured rapidly, and capabilities that were beyond the reach of a utility a decade ago are now easily accessible. Similarly, access to greater computer processing power is enabling sophisticated models to support simulation and predictive analytics. The opportunities to leverage the value of these capabilities are exciting; to capitalize on them, utilities need to shift attention to the data in the systems, and to the processes necessary to ensure data quality. The following sections will provide guidance on the considerations necessary for that.

- Data can add value beyond the initial use in an IT System.
- Data management is the set of practices involved in making the information useful for reporting and decision making.

- There are many ways of collecting data, and it is best practice to have standard procedures for data collection to ensure data quality.
- Data quality has many dimensions and is critical to achieving value from IT systems.
- Data governance is the set of roles and processes that are put in place to ensure data quality.

2.0 WHY DATA MATTERS

Much of the attention on IT in utilities is on the systems. All systems have inputs (i.e., data) and generate outputs. The system takes the data input and transforms or manipulates them to generate the outputs that provide the intended value of the system. The outputs become a new form of data. For example, a meter reading is a piece of data that contributes to a billing amount. To derive the amount, however, additional data are needed, such as the rate structure. To send the bill to receive revenue, additional data about the address are needed. If any piece of data is incorrect or out of date, the use of the outputs is diminished and the business value lost. Like a recipe, the quality of the outputs are only as good as the quality of the inputs.

Understanding the data associated with each of a utility's IT systems is necessary to build a realistic and robust IT strategy. There are many potentially high-value IT applications including digital twins, artificial intelligence, and virtual design and construction that can greatly improve a utility's ability to respond to events, predict customer needs, and streamline capital programs, if the data are available. All of these capabilities require that the data in multiple systems are of high quality and that the data sets from different systems can be combined in a manner that enables advanced analysis. Data models are key components of the architecture of all IT systems. These data models include the definition of each piece of data, and its relationship to other data elements in the system.

In addition to the specific data elements themselves, there are rules governing the data values and relationships, and metadata. Metadata are data about the data that provide context to support the appropriate use of the data. For example, for a water quality sample, the data elements will include values for the parameters measured, and metadata will include the time and location that the data are collected. Systems will build metadata and rules into their data models to facilitate data quality. This chapter discusses the primary data associated with each utility IT system.

3.0 DATA QUALITY

There are multiple dimensions to data quality. Although typically one may think of the accuracy of the data as the primary concern, there are many other things to consider. There are many data and information quality frameworks that have evolved over the years, and most references agree that the following six dimensions provide an adequate perspective: accuracy, completeness, timeliness, consistency, validity, and uniqueness.

Accuracy is a measure of the correctness and precision of the data. For an individual data element like ZIP code, it may be a simple question of if the datum is correct or not. When considering location data more broadly, it is also important to consider precision. Although a property may be accurately represented by an address, a leak in a valve will require greater accuracy. Data collection devices often come with ranges of accuracy so the consumer of the data can understand within what distance to expect to locate an item. For example, a GPS may collect data within 1 m of accuracy, but a light detection and ranging (LiDAR) device can be accurate with centimeters.

Data sets can also be evaluated for accuracy. This is particularly useful for evaluating the quality of manually entered data where errors can easily be made. The accuracy of a data set may be measured by the percentage of data elements that are correct.

Completeness—it is possible for data to be accurate but incomplete. A customer address with an accurate street name value and missing number is incomplete. Established IT systems have safeguards built into the user interface to ensure that necessary data is entered before continuing in the system. Systems will often have notations on the entry screen to indicate required fields versus optional fields. Completeness is also relative to the intended future uses of the system. For example, in a computerized maintenance management system (CMMS), an asset may have key data but be missing elements such as criticality or date of installation. In cases like this, the data may be useful for work orders, but the value for asset management would be diminished. It is important when evaluating data from completeness that you consider all of the intended uses.

Timeliness—when data describes something that changes over time, it is important to understand the timeliness of the data. This is important for financial data and water quality data, for example. If you are making financial decisions, for example, it is important to know the budget available at the current time. If revenue and expenditures are not updated or calculated with sufficient frequency, this can lead to poor financial decision making.

Similarly, when reporting data to the public and to regulatory authorities, the expectation is that the data is current. Water quality sample data, for example, must be reported with the data of sample collection for it to be of value. This is important for transparency and for public trust.

Consistency—there may be pieces of data that are stored in multiple systems. If value for the same data differs between systems, this results in inconsistency and loss of faith in the data. An example of this may be data about the location of an asset. If as-built drawings indicate a specific location but the maintenance management system has a different location, it may be difficult to know which is the accurate value. Best practice for addressing consistency is to define the system of record for each type of data, and to establish procedures for synchronizing redundant data items between systems using the system of record to keep the other systems up to date.

Validity—systems have defined rules for data formats. If data is not entered with the proper format, then it could be rejected from the system. This is often addressed through a system's user interface by preventing invalid values. For example, a phone number must conform to the (999) 999-9999 format and letters cannot be entered where numbers are expected. When a system has custom fields or configurable fields to extend data collection and analysis capabilities, it is important to define the rules for collecting that data. For example, if you are assigning a criticality score to an asset, it is important to ensure that only data value that represents your scoring methodology can be entered to ensure that data is available for use in decision making.

Uniqueness—this aspect refers to ensuring that multiple records are not entered into a system. For example, if a customer is entered into the system twice, once with one spelling and second with a typo, it's possible you could have duplicate customer data. To ensure data uniqueness, systems should require a unique identifier. For a person, that could be a Social Security number or government-issued identifier; for a fleet vehicle, that could be a vehicle identification number.

4.0 DATA SOURCES

Utilities collect, store, and generate a significant amount of data from both IT and operational technology (OT) systems such as those discussed in Chapter 2. All these systems are sources of data that can potentially be used for multiple purposes. Over the past decade, the role of data in utilities has evolved, and this evolution will continue as people experience the value of the insights obtained from combining and analyzing data sets.

There are sources of data that have not yet been fully digitized in many utilities. The physical infrastructure has been built and maintained over many decades, and as-built records are often incomplete. In some cases, there are paper documents that also serve as data sources, and in others the physical asset itself is referred to as a data source.

To be best prepared to capitalize on this emerging value, utilities need to have a complete understanding and documentation of their data sets. This could include the following:

- source system (CIS, CMMS, LIMS, GIS, SCADA, etc.)
- primary data sets managed in each system with key identifiers (customer data, asset locations, etc.)
- primary uses of data (billing, work orders, regulatory reporting)
- data sets and elements that integrate across multiple systems
- staff roles/users responsible for each data set
- measure of data quality for each data set (see measures in Section 3.0)

It can also be useful to include external data sets in this, such as rainfall data from the National Oceanic and Atmospheric Administration, roadway data from the relevant department of transportation, data about impermeable surfaces and soils, and topography data from the U.S. Geological Survey. These data sets can provide valuable context in modeling water resources and watershed behaviors.

Following this explanation of data, the next step is to identify desired business uses for the data. An organization may wish to be able to apply advanced predictive analytics to model effects of climate change, for example, or may be interested in building a digital twin to understand how assets might perform under certain conditions. To do this, there could be different measures of quality for the data needed compared to the data's intended use in their source systems. A data assessment and gap analysis will need to be done based upon the current and potential future uses of the data.

5.0 DATA COLLECTION

There are a number of data collection efforts that are standard procedure for a utility. This includes procedures to collect information on a new customer, water sampling and analysis documentation procedures, and asset identification and analysis procedures. For each procedure, there will be a set of specifications to ensure that the data collected meet all the data quality requirements. Designated staff should be trained to understand the data collection process and responsibilities for data quality should be set out in job descriptions for staff as appropriate to their role and position. For example, staff responsible for billing would be trained to collect new customer information and assign the appropriate rates, and laboratory staff could be responsible for water quality sampling data.

As new business uses for data present themselves, and as more complex data sets are considered to support decision making, utilities are recognizing that dedicated data collection efforts may be necessary. Asset management programs are an example of an enterprise program that often requires additional data collection efforts. These efforts involve defining the required and desired data elements necessary to support the program, and the development of a tool and procedure for capturing that. Handheld devices are most effective when data collection requires fieldwork. Standard data collection forms can be created and stored on the device, and staff training provided for those responsible for collecting the information. Procedures are then created to validate the data collected and flag any data of questionable quality based on the rules defined. The next step is to combine the data records into the appropriate source system database. For asset management, this is often the CMMS.

Sensors are another mechanism to collect data. The applications are evolving rapidly and include everything from flowrates and water levels to measurements of the presence of specific pollutants and viruses. The use of sensors requires proactive planning as with any data collection effort. The selection of sensors for a specific use, the placement of the sensors, data validation procedures, and sensor maintenance must all be thoroughly considered before deploying the technology. Once done, the data provided by the sensors can be a great resource for staff operating the facilities and collection systems because more data can be collected, and real-time feedback can be provided to staff if a measurement is out of line with desired values.

More advanced data collection procedures may be applied to collect physical asset data. Because digital records were not common or even available when some assets were built, and as capital programs have modified assets over time, there are often not complete records of as-built or as-exists assets. Advances in technologies such as LiDAR and 3-D laser scanning have made it possible and cost effective to capture valuable information in digital form for use in building information modeling for facility rehabilitation and construction as well as for use in CMMS for asset maintenance and management. Figure 3.1 presents an example of the results of laser scanning to support construction project planning and design.

Data collection efforts take time and resources, and it is not a place to cut corners. This sets the foundation for all future uses of the data, and data quality must be the top priority. However, before embarking on a major data collection effort, it is important to identify the staff role responsible for maintaining that data. No data is static. Therefore someone must be designated through standard procedures to keep the data up to date as

FIGURE 3.1 3D-Laser Scanning Is an Integral Part of the Building Information Modeling Process (CDM Smith) (Tamblyn, 2018)

conditions change over time. Data collection without data stewardship will result in diminished value of the efforts over time.

6.0 INTEGRATING DATA SETS

Some of the most powerful uses of data can be found by bringing together elements of different IT systems. Table 3.1 presents some common examples of system integration points between an enterprise asset management (EAM) system or CMMS and other systems to support different business needs. The governance bodies of utilities and the public are now expecting utilities to have and use their data to explain and justify their projects and expenditures. This requires data from different sources.

Predicting system behavior and risk management are additional areas where integrating data can provide useful insights. A geographic information system (GIS) is one type of IT tool that is well suited to supporting data integration. By design, the tool is intended to support multiple layers of data, especially those that can be captured and managed on a map. New data layers can be brought in to support comparison and visualization of trends and relationships. For example, a storm water risk analysis in Fairfax, Virginia,

TABLE 3.1 Points of System Integration for an EAM/CMMS

System	Integration Areas	Purpose of Integration
GIS	Asset and customer locations	Routing and work scheduling Customer inquiries Geographic trend analysis
Financial system	Asset (value), maintenance budget/costs, and capital budget/costs	Governmental Accounting Standards Board reporting Capital planning Job costing for reimbursement Inventory purchasing
CIS	Customer locations	Link service calls to work orders Call logging
Human resources	Employee data	Job costing of labor resources Work assignment based on staff skills
Operations and maintenance manuals	Equipment servicing guidelines/standards	Preventive maintenance scheduling
Process control	Equipment run times	Preventive maintenance scheduling

used GIS to look at customer complaints from the CMMS with data about soil, impervious surfaces, and tree canopies to inform the prioritization of sites where preventive maintenance should be done to prevent erosion.

Care must be taken when choosing to integrate data sets. It is important that the intention and quality of each data set be considered, and that any potential biases in the data be considered for their effect on results. This is especially true as utilities look to expand their use of data with artificial intelligence, the Internet of Things (IoT), and digital twins. These emerging areas all involve bringing together different data sets and evaluating data for appropriateness of alignment with other data elements, and for validity for the intended use is critical. For example, in looking at asset failure, it is important to consider multiple factors including age, maintenance history, environmental conditions surrounding the asset, and the like. It can be challenging to prove a causal relationship, yet the insights gained by considering these factors together can lead to a robust preventive maintenance program.

7.0 DATA GOVERNANCE

Data governance is the set of roles, responsibilities, and processes to ensure data quality. Each system has a number of procedures for data entry, and best practice is to also have procedures to ensure that data meets data quality standards as outlined in Section 3.0. As data sets begin to be combined for analysis and decision making, such as the justification of capital investments, setting policy, or predicting system behavior, it becomes more important to have defined roles and responsibilities related to data and its use.

Why is data governance important? Because utilities are spending significant amounts of money and staff time on IT systems. These systems can and should be able to support the utility in serving the public and protecting our natural resources more effectively and efficiently. This will only be the case, however, if the data in the systems is sound and trustworthy.

How do you start a data governance program? Because it can be overwhelming to consider all the data of the utility collectively, it is best to start with a specific system or business initiative. For example, asset management or climate change mitigation. Within the context of that program, you can begin by defining the goals of the system or initiative, and the role of data in meeting those goals. The next step is to evaluate the current state of data needed, and to identify any data quality gaps. These gaps can be prioritized in terms of their contribution to the overall goals, and activities can be defined to address them. Through these activities, roles and responsibilities for data quality ownership will be assigned for each relevant data set or system, and procedures will be defined to assess and maintain data quality. This will include monitoring procedures and steps to correct data that are not of adequate quality. Another aspect of data governance is ensuring that data are not taken out of context and used for a purpose for which they are not appropriate. For example, spatial data may be accurate within some spatial limits such as if manholes are identified by aerial photography for a GIS. The defined data stewards can inform potential users of that data of the spatial accuracy and its ability to meet the needs of the intended use.

The primary benefit of a data governance program is to protect the utility's investment in IT systems, activities, and infrastructure. However, this extends even further to managing risk, ensuring decisions made with data are based on sound information, and maintaining trust with stakeholders.

8.0 POTENTIAL FOR THE UTILITY OF THE FUTURE

We are at a point in history where the value of IT is shifting from the software to the data. Each IT software system has its importance to meet standard business needs such as operating the facility, maintaining assets, billing customers, and reporting to regulatory agencies. Even so, the potential to be able to address the dynamic world in which utilities operate, including effects of climate change, emerging contaminants, economic and resource volatility, and public expectations, will require more advanced analyses. Some of the emerging uses for data are described here.

8.1 The Internet of Things

At its core, IoT is the use of technology devices to perform activities that would have otherwise been done manually. It refers to devices being connected and sending data to each other to complete a task. For example, there are utilities that are putting in sensor networks in the collection system to trigger an alarm being sent to a maintenance crew or supervisory control and data acquisition (SCADA) system to prevent sanitary sewer overflows (Dunne & Turner, 2020). In this case the thing is the sensor, and it is connected by a wireless network to the internet and sends a signal to the CMMS or SCADA system when a certain condition is observed. Data are the central element here that are being shared between systems to provide real-time information that would not be feasible with manual observations.

8.2 Digital Twins

Utilities have been using models for many decades to run simulations and evaluate scenarios. The intention of a digital twin is to build on these by including greater detail in the data and validating simulation results with real-time operations to the point where the actual system and the model are providing the same results. Achieving this level of accuracy provides a valuable tool for a utility be able to predict performance of its system under different conditions to evaluate risks. Combined sewer overflows are an area where this is being applied, as presented in Figure 3.2. In this example, it is again the data that is the central focus. The comparison of simulated and actual data over time is the output of value.

8.3 Artificial Intelligence

The availability of powerful computing technologies that can process large data sets of information about water systems quickly has led to the development of artificial intelligence (AI) capabilities. Artificial intelligence refers to the analysis of interpretation of these data sets by computers to initiate action. For example, utilities in Europe are exploring the use of AI to support robotics in sewer systems to address aging infrastructure (Benjamin, 2020). Most utilities are not comfortable with the idea of automated action based on computer simulations. This is in part due to lack of trust in the data quality, lack of understanding of the analytics involved, and reluctance to override human experience with automation. There is, however, a tremendous opportunity to use AI-based tools to support human decision

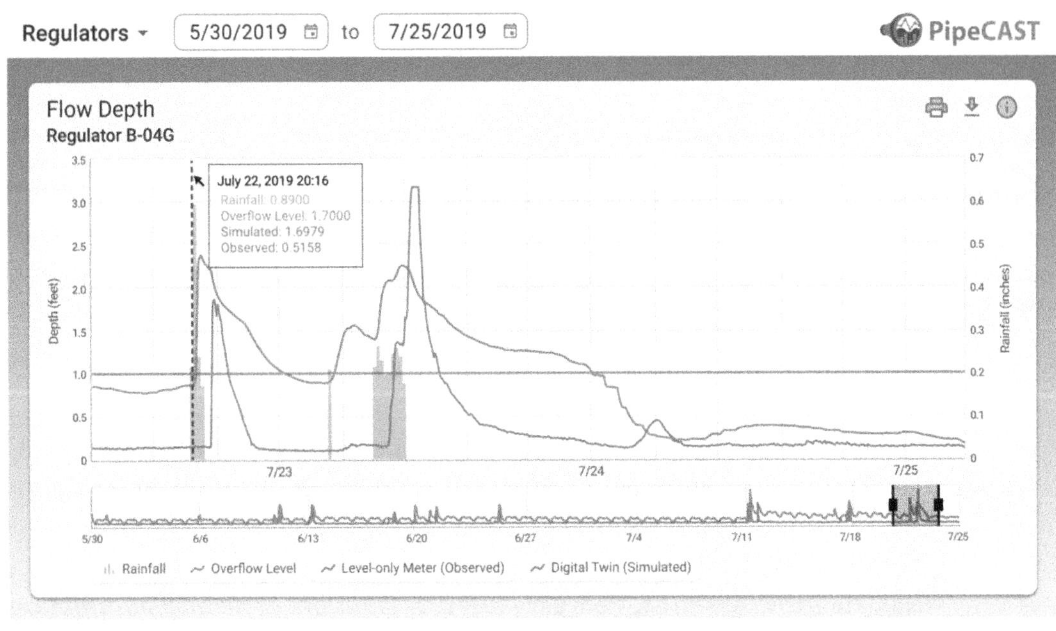

FIGURE 3.2 Digital Twin Data Presentation: Comparison of Overflow Levels Simulated by the Digital Twin Compared to Observed Values During Wet Weather (Reprinted with permission from Amy Corriveau, CDM Smith)

making and action taking as machine-assisted learning, where the AI tools analyze and interpret data sets that are too large and complex for humans to process and present the results and recommendations for staff to act on.

To achieve the potential of any of these advanced IT tools, it is imperative that staff be intimately involved. Reliance on IT for decision making requires a level of trust and understanding that develops over time and with experience of use. Staff understand the data, and can help address data quality, and they can come to realize that these tools are not meant to replace humans, but rather to help us navigate the dynamic environments in which we operate to better serve our communities.

9.0 REFERENCES

Benjamin, T. (2020, March/April). Harnessing the full benefit of AI in water. *World Water Magazine*.

Dunne, S., & Turner, V. (2020, June 2–5). *The effect of the Internet of Things (IoT) on operations and maintenance* [Conference session]. WEF Collection Systems Conference, El Paso, TX, United States.

Tamblyn, R. (2018, February). Begin with the end in mind: Modeling tools and quality data power maintenance management systems. *Water Environment & Technology*.

10.0 SUGGESTED READINGS

Caldwell, E. (2020, August). Facility maintenance transitions from reactive to proactive. *World Water: Stormwater Management Magazine*. https://www.accesswater.org/?id=-10032948&fromsearch=true#iosfirsthighlight

Maheshwari, A. (2021). *Data analytics made accessible*.

Olavsrud, T. (2021, March). Data governance: A best practices framework for managing data assets. *CIO Magazine*. https://www.cio.com/article/3521011/what-is-data-governance-a-best-practices-framework-for-managing-data-assets.html

Ramamurthy, A. (2018, February 20–23). *Data life cycle and information management: A data driven road map for enterprise decision making* [Conference session]. WEF Utility Management Conference, San Antonio, TX, United States.

Tam, B. (2020). Harnessing the full benefit of AI in water. *World Water Magazine*, *43*(2).

Von Sperling, M., Verbyla, M. E., & Oliveira, S. M. (2020). *Assessment of treatment plant performance and water quality data*. IWA Press.

4

Planning and Implementing Information Technology Projects and Programs

1.0 SUMMARY OF KEY THINGS TO KNOW

- Information technology (IT) represents a significant investment for utilities and should be considered in the context of the broader utility strategic priorities.
- IT-specific strategic priorities can be documented as part of the utility's strategic planning process, or as its own effort.
- The planning process is most effective when it is inclusive and transparent and has executive involvement.
- There are a variety of tools available to assess, document, and prioritize IT needs.
- The results of the plan should be used to inform IT expenditures and initiatives.

2.0 ALIGNMENT WITH UTILITY PRIORITIES

2.1 Relationship Between Information Technology and Utility Strategic Priorities

As discussed in earlier chapters, IT systems are most effective when they facilitate a utility's ability to meet business objectives. Prioritization of investments in IT should be aligned with the utility's business priorities. Although most utilities conduct an annual planning effort to support budgeting, it is also recommended that a longer term perspective be taken to evaluate needs over a 3- to 5-year time frame given the cost, complexity, and duration of utility capital projects and programs. The Government Finance Officers Association (GFOA) (Chicago, Illinois) recommends that all governmental entities use some form of strategic planning to provide a long-term perspective and establish logical links between authorized spending and broad organizational goals. In *Recommended Budget Practice on the Establishment of Strategic Plans*, GFOA (2005) defines strategic planning as "a comprehensive and systematic management tool designed to help organizations assess the current environment, anticipate and respond appropriately to changes in the environment, envision the future, increase effectiveness, develop commitment to the organization's mission and achieve consensus on strategies and objectives for achieving that mission."

The strategic planning process incorporates perspectives of multiple individuals and groups to ensure that the planning process includes landscape element analysis, organization element analysis, and adaptive planning. Landscape element analysis seeks to identify external trends and drivers, such as major businesses entering or leaving the service area, environmental changes in resource supply or quality, changing customer expectations, and looming regulatory requirements. Organization element analysis focuses on internal capabilities and resources, including existing strengths and weaknesses, upcoming changes, and emerging opportunities. Adaptive planning categorizes each of these changes as orderly, dynamic, or chaotic.

A strategic plan typically identifies a limited number of high-level strategies. An action plan defines how these strategies, or goals, will be implemented. Each strategy will have one or more specific, measurable objectives. The strategic plan will also identify ongoing performance measures, sometimes in the form of a balanced scorecard, to help the organization track performance over time.

2.2 Do You Need an Information Technology Strategic Plan?

An IT strategic plan can seem like an effort and expense that cannot be afforded given a utility's many competing priorities. In fact, there are circumstances in which that is the case. Consider these questions:

- Have you recently completed a utility strategic plan or will you be completing one soon? If so, it would be prudent to extract IT needs from that. If you have a strategy, it could be possible to extract IT-related priorities from that if that was not already done, and then overlay that with IT-specific assessment areas like upgrade needs and cyber risks. If the process is imminent, the scope should be defined to also capture supporting IT priorities.
- Do you have time-sensitive IT issues to address? This could include replacing obsolete software, as well as addressing cybersecurity risks or IT needs related to regulatory reporting or other outside requirements. If this is the case, it is important not to postpone urgent activities in order to complete a strategic plan, just as you would not postpone a water main repair until you complete your capital planning process.
- Have you ever done an IT strategic plan? If so, consider how that plan was used, and if its content and context are still relevant today. It may be feasible to update the plan rather than start from scratch.

- Are your IT systems serving you, and do you have a good understanding of upcoming IT expenditures and their value to the utility and your stakeholders? If not, some effort should be spent to provide the perspective necessary to ensure that your IT systems are serving the needs of the organization.

Information technology strategic planning is best performed within the context of an organizational strategic plan. In this scenario, IT investments can be linked to organizational goals, objectives, and measures. A certain level of strategic analysis is vital within the IT planning process to ensure that IT activities anticipate, leverage, and support changes affecting the utility in a way that best supports the utility's long-term success.

In general, IT strategic plans will identify a series of programs, projects, and action items for implementation over the period of the plan. Many of the activities used in creation of a strategic plan are repeated within the context of each program and project. The difference lies both in the breadth and depth of application. Strategy covers the entire organization, but at a high level. Programs cover a specific set of projects and initiatives that are related in some manner. For example, an asset management program might involve implementation of a work management system, data collection activities, and interfaces between various systems. Each of these might be structured as a separate project, or efforts can be organized as a program to highlight the interrelationships between disparate systems and business processes. Chapter 6 discusses the processes involved in program management.

An IT strategic plan is best viewed as a project, with a specific starting point (typically a project kickoff meeting) and a specific outcome (i.e., the published plan). This chapter presents the participants, methods, techniques, and deliverables commonly used in the creation of a strategic IT master plan. Business objectives of the plan are identified, and an overview is provided for the plan creation process. Table 4.1 provides an overview of the processes, tools, participants, and deliverables discussed in this chapter. In addition, the case study presented in Section 2.0 in Chapter 9 provides an example of one utility used their strategic plan to guide their IT priorities.

3.0 STEPS TO UNDERSTAND YOUR INFORMATION TECHNOLOGY NEEDS

Begin with the end in mind. The primary goal of an IT strategic plan is to ensure that IT investments are applied in a way that makes business sense across the entire organization. The primary output of the strategic IT plan is a prioritized list of IT-related projects with budgets to include in operational

TABLE 4.1 Project Planning Overview

Process	Tools	I/W/S	Participants	Deliverables
Utility strategic planning	SWOT vision	I/W/S	EST	Drivers and objectives
Initiation and kickoff	Vision	W	EST/PT	
Identify business drivers	SWOT	I/W	EST/PT	Drivers and objectives
Review current situation – Business process mapping – Skills mapping		I/W/S	PT/IT/SMEs	
Identification of gaps	Gap analysis	W	PT	Gaps
– Disaster recovery/ business		I/W	IT	
– Continuity		I/W/S	PT/IT/SMEs	
– Service catalog/ SLAs				
Alternatives analysis	Alignment CSFs and KPIs	W	EST/PT	Prioritized opportunities, strategic direction
Plan presentation		W	EST/PT	Program definition, master schedule, and budget estimates

Note. I = interviews; W = workshops; S = surveys; EST = executive steering team; PT = planning team; IT = information technology employee group; SLAs = service-level agreements; SMEs = subject-matter experts; CSFs = critical success factors; KPIs = key performance indicators

and capital improvement plan budgeting. It is important to identify budget deadlines at the beginning of the planning process, to understand existing allocated funds, and to leverage any existing project justification documentation. The final plan should include projects that are aligned with budget cycles.

Although there are a variety of ways to carry out a strategic IT plan, typical components include identification of business drivers, current situation review, identification of gaps and alternatives analysis, project identification and budgeting, and final plan presentation. The Water Environment Federation (WEF) and the Water Research Foundation (WRF) have been supporting a more structured and systematic approach to understanding and documenting the people, process, and technology components within utilities and have created the Water Intrapreneurs for Successful Enterprises (WISE) approach (see Figure 4.1). "The goal of WISE is to leverage a systemic approach to improve different aspects of managing a water sector utility. The WISE mission is to apply system thinking and provide a methodology for utilities to improve their capabilities and enable management practices focused on value and overall performance" (Vitasovic et al., 2021).

This approach provides a thorough methodology to evaluate and plan utility needs and priorities, including IT.

3.1 Identify and Convene Key Stakeholders

Sustained executive commitment is required to successfully implement IT projects, so it is critical to have a designated executive from within the

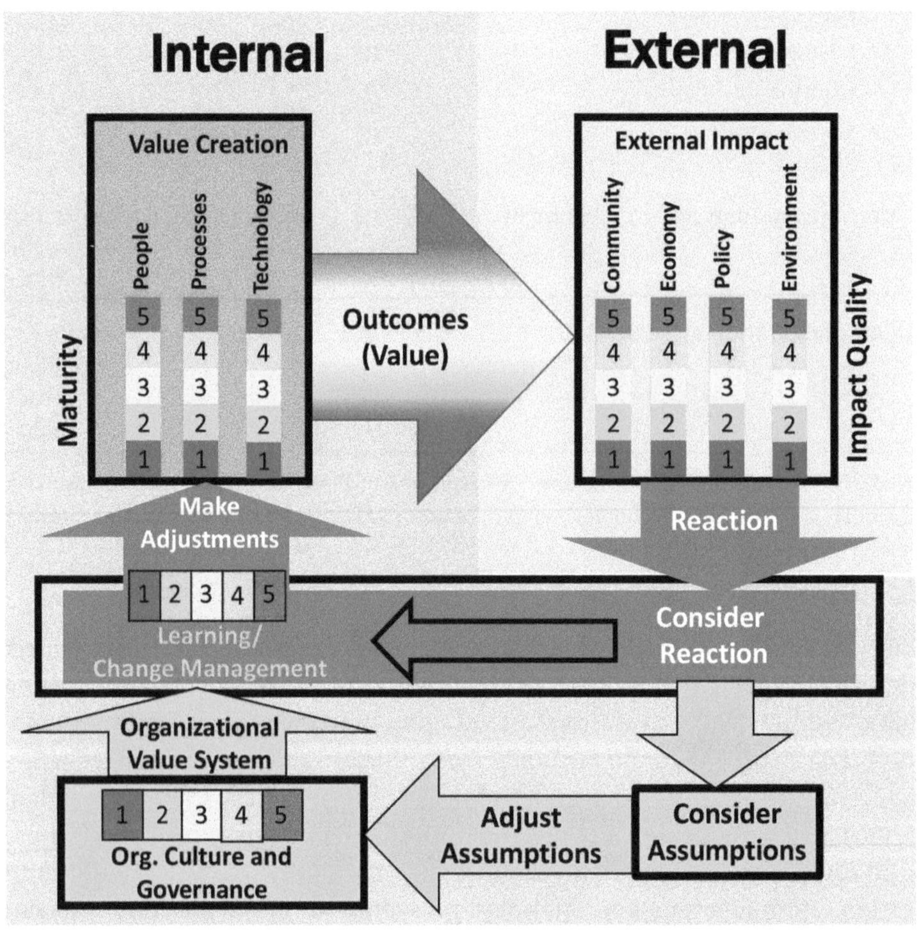

FIGURE 4.1 VCI Framework for WISE (Vitasovic et al., 2021)

utility as a champion and participant in any IT effort. This person can also help identify and ensure the support of key stakeholders to set the stage for successful outcomes. If a consultant will be hired to support the project, executive support will be required to facilitate the budgeting and procurement for that scope of work.

There must be a designated leader of the strategic planning effort from within the utility as the project manager. This person may be supported by an outside consultant, or they may be responsible for coordinating the effort with an internal team. If the effort is to be completed internally, staff must be given adequate support to temporarily reassign their work responsibilities while the project is underway.

Strategic plans are typically carried out using two small ongoing groups to provide direction; these are the executive steering team and the project team. A wide variety of individuals in more limited engagements provide depth and breadth to the planning effort. The leader of the effort need not be an IT expert. It is most important that they are able to organize, collaborate, and communicate well with diverse groups.

3.1.1 Executive Steering Team

A steering team is critical to the success of an IT plan. The steering team should include the plan project manager; the senior IT leader; and senior representatives from operations, engineering, finance, and customer service. In a large organization, additional participants may be appropriate to address the many different business units.

The executive steering team establishes the high-level vision for the organization and provides guidance to align IT priorities with business priorities. The steering team must be able to provide executive insight and direction during development of the IT plan and provide the governance support necessary to finalize a plan so that it can be included in budgeting efforts and implemented. To ensure this, individuals with a good understanding of the process and politics needed for plan approval and success should be sought after.

3.1.2 Planning Team

The planning team includes supervisors and potential end users from throughout the organization who can grasp the potential of the projects proposed, not only for their own department but also for other departments. Members should be subject-matter experts to provide knowledge about specific areas, including applications, regulations, and business processes. The IT group should be represented but should not direct the effort or dominate the membership of this group.

The planning team will contribute to the completion of deliverables and participate in analyses of alternatives and prioritization portions of the project. The team will help carry out information-gathering activities and will communicate plan progress to their respective departments. The time commitment required from members will be dependent on the scope of the project and the level of outside consulting support.

3.2 Articulate Current Conditions

It is important to compile information and perspectives that describe the current state of IT within the organization. The following resources are vital to the planning process and should be consolidated for easy access by the project team:

- annual operations budget
- most recent capital improvement plan
- most recent organizational chart, updated to reflect current organization
- most recent IT software and hardware inventory, including the date of last upgrade
- most recent IT architecture diagrams, updated to reflect current implementation
- most recent system maintenance costs, including hardware and software maintenance and fees for recurring services such as leased data or fiber
- any risk assessment studies
- the organization's strategic plan, mission, and vision

This information can be used to understand the current IT budget in total and as a percent of total budget, the number of systems and the number of users across the organization, and any known risks.

3.2.1 Business Process Modeling

In addition to facts about costs, systems, and users, it is important to set the business context and to identify functions and processes that include IT systems to support them. Business processes should be documented in their current state. If possible, capture measurements to quantify the issues, such as the time required to complete one iteration of the process and the number of times per period the process is repeated. The goal is to fully understand the current needs and requirements so that not only the "what" but also the "why" is clear to the planning team. There are a number of methodologies to do this including the WISE approach mentioned earlier.

Business process modeling represents the work done by utility staff and the interactions between their activities. In the sense that business models can be constructed from fundamental building blocks, just like models of water or wastewater treatment processes are constructed from chemical, biological, and physical components, business process modeling constructs from business tasks, workflows, resources, dependencies, and measurements. The building blocks for business process modeling are

- tasks, or activities, combined in a flow of work or process;
- decision points, including any branching of outcomes dependent on a decision;
- work objects, including tangible (e.g., product) and intangible (e.g., service) inputs and outputs;
- business rules that represent how decisions are made;
- resources required for a task (these can be either machine or human); and
- metrics that define performance for a task.

Although business process modeling has been in use to support IT planning for decades, most utilities do not maintain their process models; this effort may need to be done or updated as part of the IT plan development. This should include a high-level definition of key business processes, the systems, the stakeholders, and the end users for each. If a process is known to be in need of IT support, it can be articulated in more detail to better identify the root of the improvement need. In some cases, non-IT activities such as training, communications, process modifications, or data sharing can lead to improvement without significant IT investment.

The importance of business process modeling to IT planning is to clearly document the role of IT systems in supporting the business and to provide insight into areas where IT investment will improve business performance. Selecting metrics for a business process is important if process effectiveness and efficiency are to be continuously improved. The following are guidelines for creating effective metrics.

- Define the metrics that are currently used. At a minimum, these have been proven useful in the past.
- Create cycle times around processes that take the most time or have the most complaints and thus are likely bottlenecks. Similarly, for

costs, define metrics for the activities and resources that are most expensive.

- Define metrics that have business relevance and are not focused on IT specifics alone.

3.2.2 Strengths, Weaknesses, Opportunities, and Threats Analysis

The concept of strengths, weaknesses, opportunities, and threats (SWOT) is illustrated using a quadrant as shown in Figure 4.2. Strengths and weaknesses are internal to the organization, and opportunities and threats are external.

The context and boundaries for the analysis can be segmented by area of the business to ensure a thorough review. For example, the SWOT can be done by business area by framing the question as "What are the strengths (weaknesses, opportunities, threats) related to the use of IT in engineering? In serving customers? In meeting regulatory requirements?" And so forth.

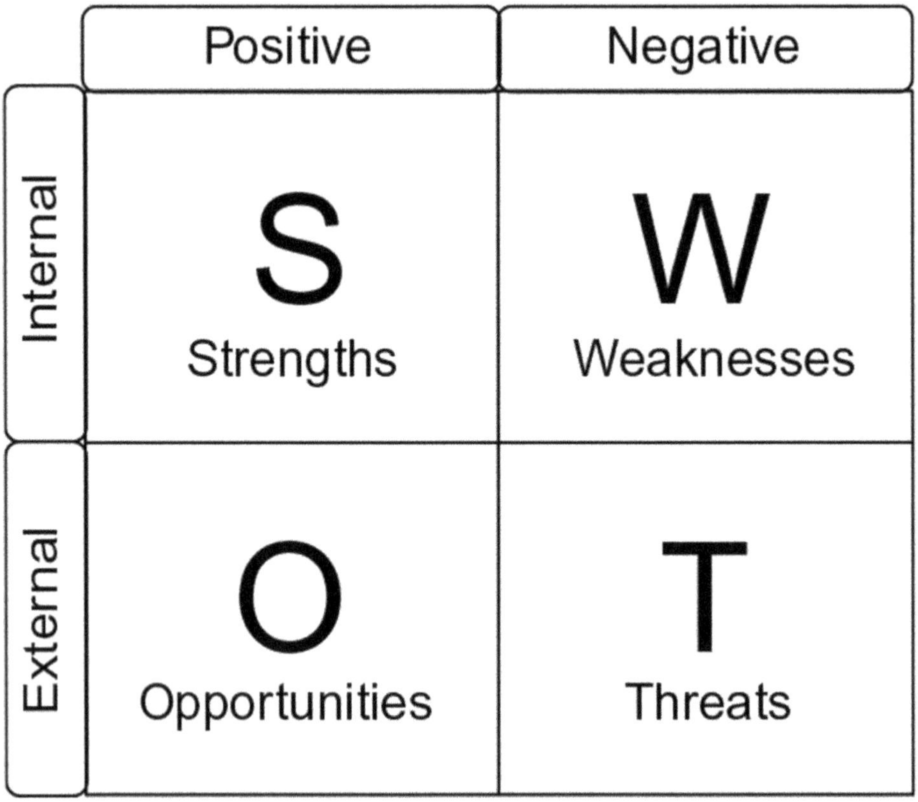

	Positive	Negative
Internal	S Strengths	W Weaknesses
External	O Opportunities	T Threats

FIGURE 4.2 Strengths, Weaknesses, Opportunities, and Threats Analysis Quadrant

The length of time required for a SWOT analysis is relative to the size of the group, the level of complexity of the external and internal environments, the skills of the people involved in data collection and analysis, and their understanding of the issues being presented. A small group might conduct a SWOT analysis in an hour or two, while a larger group might spend a half day on the exercise. The number of people involved is less important than the skill sets of the individuals involved. Consider including executive steering committee members, regulation specialists, public relations specialists, and consultants.

When the analysis is complete, the information can be compiled and consolidated by looking for themes and focusing on the areas with the greatest effect. The output will include the detailed SWOT elements, as well as an assessment of the actions that can be taken to build on strengths, address important weaknesses, capitalize on opportunities, and mitigate effective threats.

3.2.3 Perspective Gathering

As analytical as IT systems and processes may be, their use and effectiveness is subjective and influenced by human behavior and perspectives. Therefore, it is prudent to include conversations with staff and other key stakeholders to gain a more complete understanding of the current state, and people's perspectives on future potential. The three methods for this are interviews, workshops or focus groups, and surveys.

Interviews provide an extended time period for individuals to provide broad input to the planning process. Interviews with executive management provide insight into orientations toward IT and help identify critical factors for evaluating potential projects. Key customers can provide an outside perspective on critical public-facing processes. Interviews are an effective way to investigate detailed issues raised in other meetings.

Consider having utility managers conduct stakeholder interviews across department boundaries. This approach will broaden managers' perspectives on how data and technology are used in other departments and change the dynamics of the conversation compared to an interview conducted by a superior of a subordinate (e.g., from a problem-solving discussion to a broader discussion of general issues and possibilities). It is important that interviews be small and safe to elicit honest input. It can be helpful to have them be confidential if there are sensitive topics brought up.

It is often not feasible to schedule and conduct interviews with all the people who have important insights. Workshops are a proven method of quickly building a broad, robust understanding of a specific set of issues. Workshop organizers should strive to ensure that the workshop provides business value commensurate with the time commitment required by all the participants. The organizer and participants should have a clear understanding of the outcome of each workshop. Workshops can be organized around

- business needs;
- process flow mapping for key, cumbersome, or error-prone processes; and/or
- specific IT systems.

In a large organization, it can be effective to use surveys to ask a limited number of questions of a greater number of individuals. Surveys can help build data to support plan proposals if the questions are multiple choice versus essay. Essay-type questions require significantly more time to complete and should be avoided wherever possible. Weigh the value of each question against the cost to the organization of having all responders answer the question. Ask what difference it will make if the answer to each question is known, and delete questions that do not have the potential to clarify issues, support initiatives, or provide other specific information.

3.3 Identify Desired State

Information collected in current situation review, vision, interviews, and workshops reveals "gaps" between where the organization is today and where the organization desires to be in the future. Most gaps represent opportunities for improvement or risks that need to be addressed. It is important to document all gaps, at least briefly, even those that clearly cannot be addressed during the current planning cycle. Changing technologies and other factors may provide unexpected opportunities that might be missed if the need was not documented.

Strategic planning is based on the assumption that change has occurred and will continue to occur. Business drivers represent types of changes for which some action is required by IT. Here are some examples:

- Business and organizational changes include discrete events such as turnover of key staff, external events that change the expectations placed on the organization. Executives and managers are typically the best source of information regarding business changes.
- Regulatory changes can occur at the local, regional, state, and federal level. Contaminants of emerging concern are examples of current regulatory changes that can affect all utilities. Managers, the safety compliance officer, and laboratory staff are the prime sources of knowledge in these areas.

- Desire for performance improvements include efforts to identify cost savings, energy efficiency, reduced environmental effect, environmental and social justice, and customer satisfaction. Information technology can play a role in supporting all these areas by providing data and analytics to guide decision making.
- Technology changes frequently, and utilities need to invest in staying up to date with versions of software and hardware to maintain business continuity. In addition, the standard of practice in IT in utilities changes as utilities begin to adopt new tools and share their success stories with others. For example, automated meter reading was not a standard practice when this Manual of Practice was originally written, but it is today.
- Risk identification drives change. Examples of these risks include cybersecurity and climate change. Cybersecurity is addressed in detail in Chapter 8, but generally speaking it is important for the IT strategic plan to identify the activities needed to better understand and mitigate these risks to the extent possible.

To be most effective in engaging the broader business, the desired state should be framed around business drivers and the role of IT in achieving organizational goals.

3.4 Prioritize Needs and Opportunities

The organization of how to present the needs associated with a desired state must be aligned with business priorities. The executive steering committee can be helpful in determining how best to document and structure this information. A long list of wants will not be as likely to get funding or support as a well-thought-out explanation of why a proposed action or investment will improve business outcomes. The organization of needs should be aligned with the structure, language, and culture of the utility, for example:

- defining and ranking IT needs as improvement opportunities or risk mitigations
- organizing IT needs by departmental business functions
- organizing IT needs in alignment with business drivers or organizational initiatives

It is often the case that there is more than one approach to meeting a need, so presenting alternatives will allow for more robust evaluations and better informed decisions. The "do nothing" alternative should be included in every evaluation and should include an estimated range of life expectancy. For example, there are times when a software system must be replaced due to obsolescence or because it is being discontinued by the vendor. The cost and effort involved may not be aligned with other utility initiatives, but the cost of doing nothing can put the utility at risk. In this case, highlighting those risks and the anticipated business effect provides the context for informed planning.

Small needs and opportunities may have one clearly defined option. Larger needs and opportunities merit more analysis and may have three or more alternatives worthy of investigation. For example, customer information system alternatives might include do nothing, minor upgrade with current vendor, major upgrade with current vendor, move to new product with current vendor, and full procurement process for new vendor/application.

Each alternative should have an identified range of budget, time, and support requirements. Ongoing support costs and interim staffing requirements should be identified where appropriate. The goal, at this point, is to understand the issues without selecting a definitive path forward.

Selecting which projects to approve and an alternative for each project should be based on clearly defined and supportable prioritization criteria. All recommended projects should be analyzed using the same criteria, although not all criteria will apply to all projects. Prioritization affects not only what projects are approved, but also the implementation sequence. Prerequisites should inherit the prioritization factors of their dependent projects.

Prioritization factors should include utility strategies, if available, and costs, value, and risks. Factors should be weighted; some factors, such as the ability of the utility to carry out the project, may have a go/no-go effect on the evaluation process. If a prioritization process reveals a ranking that seems wrong to the project team, prioritization criteria should be reevaluated to identify missing factors. The goal is to clearly define and document the actual prioritization process.

Important information to include in the prioritization process includes the following:

- Organizational effects—Staffing effects of projects under consideration should be identified. Effects to IT and department staff in the areas where the new technology will be implemented should be included. Backfill and temporary staffing potential should be evaluated. The roles of consultants should be defined to provide specialty skills, tasks that are repeated only once every several years (such as this planning process), and in areas where the utility does not have core competency in-house.

- Budget—It is important for the designer to understand the availability of funding over the plan period. In addition to understanding the potential annual budget and capital improvement funds available, some funding may be tied to other, non-IT projects, while other funding may be department and/or project specific.
- External factors—Upper management and the public relations officer can provide input on local political considerations that may affect project timing and approval. Upper management and regulation specialists can identify political factors on a local and federal level that may increase the priority of specific projects.

4.0 DOCUMENT THE STRATEGIC INFORMATION TECHNOLOGY PLAN

As important as gathering the needs and priorities is the process of documenting them. A strategic IT plan should be written to provide a useful road map for years, and can be used to track and monitor progress. The content of the plan should include the business context for the efforts and resulting recommendations as well as the projects and programs needed to support business priorities, The plan should set forth appropriate expectations of project time, expense, and staffing support requirements. The plan should carefully document the desired benefits to be provided by the project, and each project should be designed in a manner that identifies the business value of each phase. To maintain executive commitment, these business benefits should be tracked and communicated when achieved.

The content of an IT program plan can vary widely. Although no strict guideline exists, the following is a generic outline that can be expanding or trimmed, as necessary:

- Approval/signature page
- Executive summary
- Discussion of the SWOT analysis
- Discussion of objectives and requirements to be met and why
- Discussion of any constraints or limitations
- Discussion of gap analysis
- Discussion of options evaluated and selected to fill the gap
- Strategic discussion on how to get from here to there (i.e., what needs to happen):

  - Organizational effects
  - Business process effects
  - Technical effects
- Tactical plan on how to get from here to there (i.e., how it needs to happen):
  - Specific changes to the organization
  - Specific changes to detailed business processes
  - Critical success criteria
  - List of specific projects, with schedule, resource, and budget requirements
  - Discussion of interrelationship and links between projects
  - Summary schedule, resource plan, and budget
- Risk analysis and mitigation
- Business case summary
- Next steps
- Conclusion

The aforementioned outline is relatively simple and self-explanatory. Whereas the specifics may vary from organization to organization, it is a straightforward layout of information that should walk interested parties through from a problem definition to a logical solution and leave readers with a clear sense of understanding and resulting support. The program plan should be easy to read, in lay terms, and with a business sense to it. Technology, as described in the plan, is presented as a *derived* response to business needs, and should also be presented in lay terms as well. Conversely, it is not recommended that the program plan be a dissertation on hardware, software, networks, and so forth; these details should be handled elsewhere, perhaps as an appendix or separate technical study for those interested in the technology details. Such technical details should, however, be ready as backup that there is a solid foundation to the final recommendations.

4.1 Business Context

An IT plan will be better focused and better received if it is aligned with an existing organizational vision, goals, and objectives. The strategic IT plan document should include these elements as they were defined during the planning process.

4.1.1 Vision

The vision should be articulated at the outset of the planning process. A vision statement should be short and clear. It can be difficult to write a concise, meaningful vision statement. Longer statements are easier but may not yield the benefits in alignment of employee behavior that are possible when a short, clear vision is well communicated.

To start the development of a vision statement, list words important to the organization in the delivery of IT services. Look for redundancy and cut out words with the least effect. If the team contributes sentences, capture them all then look within the sentences for key words and phrases. If a short, clear vision statement cannot be obtained within a reasonable amount of time, move ahead with the shortest version that receives approval by the group.

4.1.2 Goals

Goals should be actionable and business related. Framing the goals with the business drivers provides a clear understanding about the role of the IT activity to support the utility. Goals may be as short as a few words and should be no more than two sentences each. There is no ideal number of goals, but more than a dozen will can overwhelm and dilute the importance of each.

4.1.3 Objectives

Objectives should be measurable proof that goals have been met. The steering committee will have both broad and specific goals at the beginning of the IT planning project. Working with the steering team to quantify these goals will increase the planning team's understanding of the goals and increase the possibility of the plan achieving these goals.

4.2 Program and Project Recommendations

Recommendations will be presented in a prioritized manner based upon the business drivers and decision logic decided upon by the team. The logic and context should be briefly explained so that people who were not involved in the planning process and those who refer to it over time can understand the reasons for the recommendations. If a project is proposed to address a gap or because it represents an improvement opportunity, this should be explained along with any relevant information about the necessary timing.

The document should not include detailed project plans. At this point, projects will be defined to the extent necessary to explain the need, the conceptual scope, the intended outcomes, and an estimated range for the timing and resource needs. The intent of these recommendations is to identify a sponsor and obtain authorization to undertake the project. Detailed project scoping and planning will occur after project authorization is given. For large scopes of work with many components that will require more than a year to implement, it may be preferable to organize projects as part of a program. Enterprise resource planning (ERP) and asset management implementations are examples of this where many different departments will be involved, each with their own scope that contributes to broader program goals.

4.3 Timing and Resources

The IT planning process is most effective when conducted regularly and as part of business planning. When a utility is considering a significant shift related to IT, it can be helpful to complete an IT-specific strategic plan. When this is done it should be updated as major projects are initiated and completed, at least annually. The plan will include a prioritized list of projects based upon an understanding of the staffing, budget, and political factors influencing project timing; each project should be evaluated for project phasing. Some projects will have clear requirements, selection, and implementation phases, while others will proceed straight to implementation.

To understand the feasibility of completing the suite of recommendations, resource needs for each project must be understood. This can be done based upon prior experience of staff, or with input from vendors, consultants, and colleagues at peer utilities. Defining the number of full-time employees needed from each department, or by area of expertise, over the course of the project will allow evaluators of the needs to understand the level of effect the project will have on staff workloads and the need to bring in external staff to support the effort. This information can also help in selecting from different alternatives if some have less of a demand for staff time. For example, many utilities are now shifting key business systems to a vendor-hosted and managed environment to reduce the demands on utility or city IT resources.

Not all strategic IT plans include schedules. This is based upon the requirements from the executive steering team and/or the ultimate decision-making entity. At the very least, however, projects with critical time considerations should be highlighted with the necessary completion date documented. For example, if a system will become unsupported by a vendor on a specific date or if a need is driven by a regulatory deadline, this should be made clear.

5.0 CRITICAL SUCCESS FACTORS AND KEY PERFORMANCE INDICATORS

Critical success factors define key areas of performance that are essential for the organization to accomplish its mission. Critical success factors are general in nature and most are industry specific. Key performance indicators are measurable and help management gauge organization effectiveness in support of critical success factors. Critical success factors can be identified at an organizational, departmental, and project level. This discussion focuses on critical success factors for the strategic IT plan.

5.1 Business Metrics

Factors supporting approval of a project are good launch points for the development of critical success factors for an IT project. Analysis performed during the planning project can help identify key performance indicators that the organization expects to improve as a result of the project. If possible, key performance indicators should be put in place before the project so that a before-and-after perspective can be obtained.

5.2 Executive Involvement

It is important to identify "measurements that matter" to the organization's management. What will management do based on these numbers? It is wise to consider how supervisors will investigate issues that are brought to the attention of management.

5.3 Focus on Implementation of the Plan

The intention of the strategic IT plan is to outline priorities to guide future IT investments and activities. The needs and recommendations all represent actions that will be implemented in the coming 1 to 5 years and should be described in such a way to jump-start the detailed project scoping process that will occur when approval and funding are granted. By beginning with the end in mind, the planning process and final documentation provide a critical foundation for long-term understanding and success.

6.0 STRATEGIC INFORMATION TECHNOLOGY PLAN OPTIONAL CONTENT

6.1 Disaster Recovery and Business Continuity

Disaster recovery and business continuity are some of the most important concepts that must be addressed in a utility today and should be included at some level in the strategic planning process. The level of detail needed for a thorough assessment, however, is often beyond the scope of a strategic IT plan. Depending on the utility's past efforts in this area, the IT plan could either reference recent risk assessments and findings related to IT, or could put as a high-priority recommendation to complete such an assessment.

A full disaster recovery and business continuity plan should identify the utility's critical functions and personnel and ensure that data, applications, and workstations will be available.

A full disaster recovery plan should include the following steps:

- Identify and prioritize IT applications, networks, and other services.
- Which IT assets will need to be recovered right away versus those whose recovery could be delayed for days?
- What is the effect on staff, customers, and the environment if systems are not working for an extended period?
- What backup systems could be used (such as non-network-dependent laptops, paper and pencil recording, etc.) if the network is down for an extended period?
- How might staff access critical databases if phone, internet, and transportation networks are not usable?

The urgency for this type of planning may be area specific, but all agencies should have this type of plan, which can be implemented when needed.

6.2 Service Catalogs and Service-Level Agreements

A service catalog, at the most basic level, is a list of services provided by an entity. A service-level agreement (SLA) is an agreement between two parties on the level of services to be provided, communication protocols in case of service interruptions, and escalation procedures in case of inadequate performance by either party. Service-level agreements are intended to increase understanding by the service provider of the consequences of service interruptions and by the customer (i.e., entity receiving service) on the alternatives for reducing or mitigating the consequences of significant service interruptions and the cost implications of those alternatives.

Service-level agreements are an effective tool for use by IT groups in establishing and maintaining priorities during service interruption incidents. At a minimum, SLAs should cover the following topics:

- duration of agreement
- clear identification of parties involved in the agreement, including primary and backup contact information for each party
- services covered by the agreement
- normal service levels

- service levels at which contingency plans and/or penalties start to accrue
- escalation protocols and contact information

A service catalog is a reasonable precursor to the creation of an SLA. In some organizations, the service catalog serves as a living document, providing consolidated information on available services, primary and backup support providers, and other key information necessary to facilitate problem resolution.

Service-level agreements are an effective tool in establishing business need for resources, including infrastructure and staffing, as they identify business effects of service outages and the costs associated with more reliable service levels.

6.3 Select Business Process Mapping

Business process mapping is an effective tool for identifying opportunities for business process improvement based on implementation of technology tools. Flowcharts illustrate processes, decision points, and outputs of the business practice being evaluated. Various methodologies exist. The most common form of flowchart is shown in Figure 4.3. Small circles or ovals represent the start and stop of the process and are optional. Rectangles represent discrete steps or processes. Diamonds represent decision points. Parallelograms represent input/output. Arrows connect the other elements to show flow through the process. A swim-lane process flow adds rows or columns (at the author's discretion) to add an understanding of the groups or individuals responsible for each step in the process. The flowchart in Figure 4.3 is redrawn as a swim lane in Figure 4.4.

The benefit of a flowchart is that it provides a picture of process complexity, thereby facilitating identification of opportunities for process improvement. Quality experts recommend ongoing mapping of business processes, both for documentation on how to perform a task in a standard manner and to facilitate ongoing process improvements. Flowcharts can be created to document existing processes and modified to show "to be" processes based on proposed IT investments.

6.4 Skills Mapping

Skills mapping involves identifying the skill sets of existing personnel with current and projected skill requirements. This is especially relevant, for example, if the strategic IT plan proposes changing key business processes by introducing new or different tools. Skills mapping can also be performed on a broader scale to identify overall training needs within the organization.

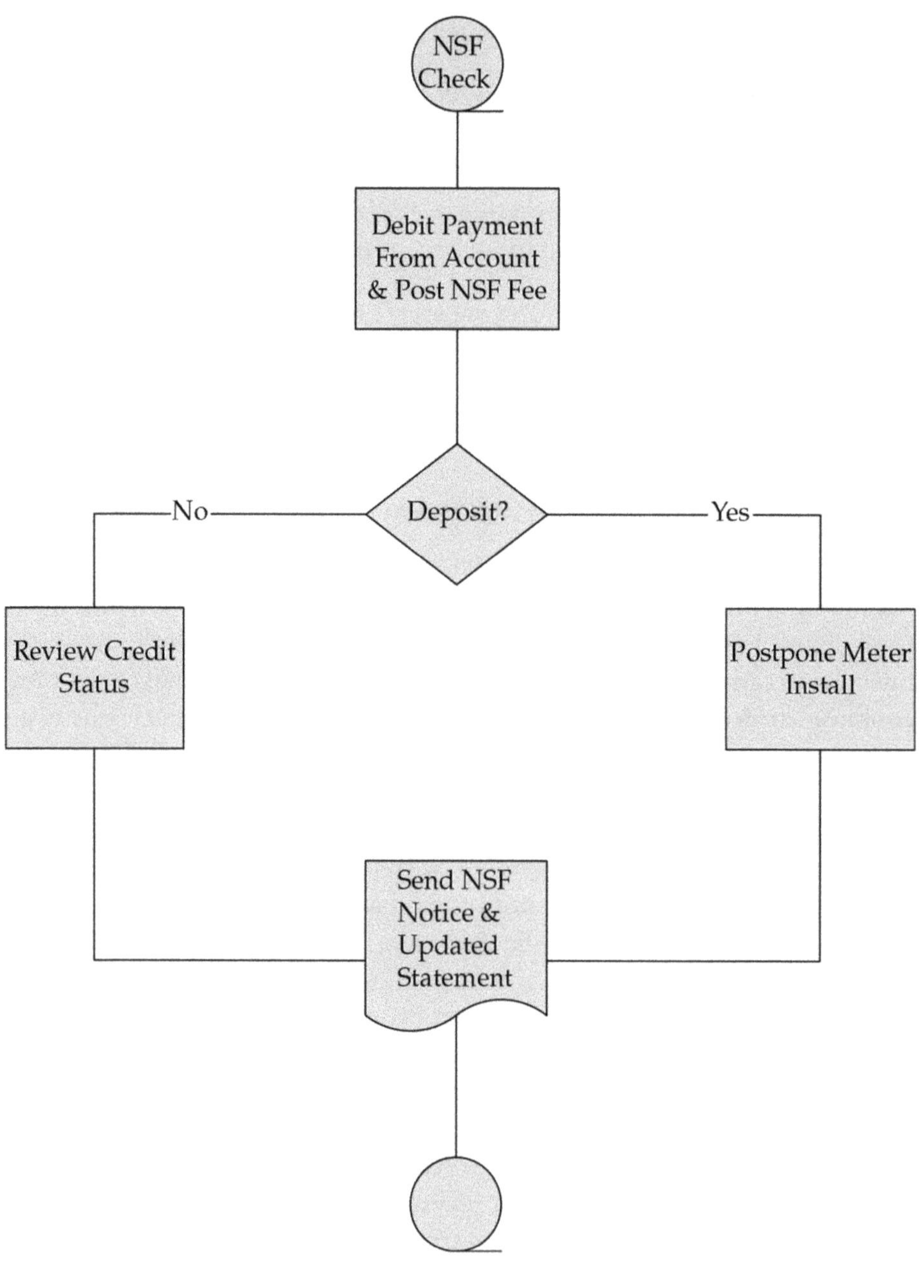

FIGURE 4.3 Business Process Mapping Flowchart

Note. NSF = non-sufficient funds

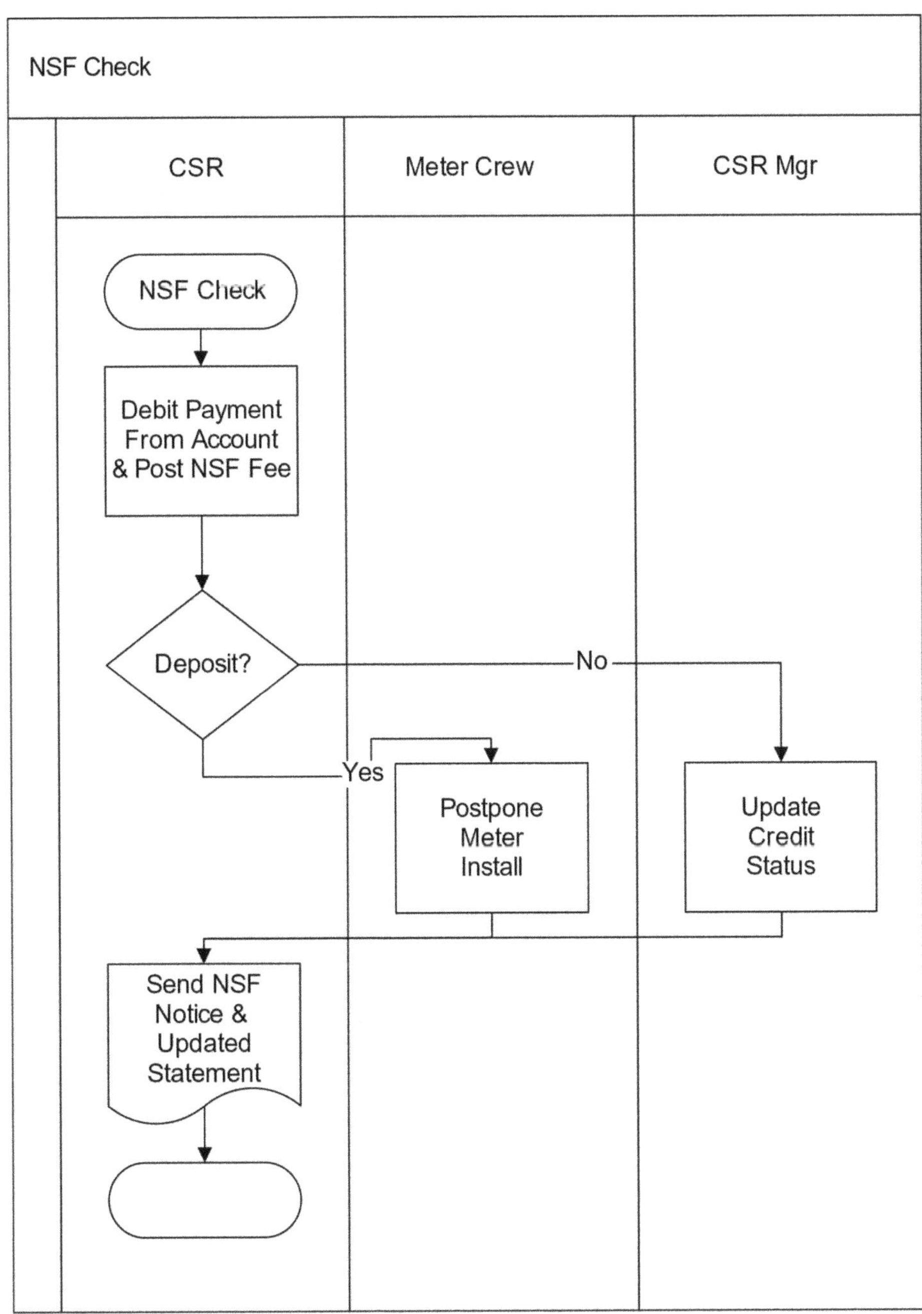

FIGURE 4.4 Swim-Lane Process Mapping Flowchart

Note. CSR = customer service representative; NSF = non-sufficient funds

For example, business process improvements are enabled in many organizations by having an appropriate distribution of business users with in-depth understanding of their primary applications. It is also valuable to measure the ability of novice users to have first-level technology questions answered by a nearby business user. In many cases, these questions reveal a lack of understanding of business processes, creating a broader learning opportunity that a remote help desk might miss.

6.5 After the Plan

In *Recommended Budget Practice on the Establishment of Strategic Plans,* GFOA (2005) calls for yearly review and adjustment of the overall IT strategic plan. Current programs, projects, and initiatives should be compared to the original goals to ensure that anticipated results remain possible based on current program and project status. In some instances, industry and/or technology changes can dramatically shift priorities, justifying a significant change in direction. In other instances, however, gentle course adjustments have unexpected consequences that can be remedied if the original goals are reviewed on a periodic basis.

7.0 IMPLEMENTING THE PLAN

The strategic IT plan should be used to guide IT-related activities and provide clarity and focus for those responsible for the utility's systems. There is not a standard approach to implementing such a plan, but there should be clear assignment of responsibility for doing so. If there is not a clear delineation of IT responsibilities and a defined IT program, it is likely that this will be a key recommendation from the strategic plan. Best practice is to manage the plan as a living document by updating it regularly based on activities completed and changing circumstances. Considering the following factors will help determine how and where energy will need to be spent. Understanding these factors will aid in developing the content of the program plan document itself, without which the content of the plan may omit key decision points. Many of these factors have been shown to be critical to program success and/or failure. As an introduction to concepts, these issues include

- executive support;
- user involvement;
- clear objectives and requirements;

- project management;
- scope control;
- drivers and constraints for status quo and desired future;
- change expectations;
- business environment and processes;
- stakeholder engagement;
- managing communications, expectations, motivation, and change;
- business case and justification (both qualitative and quantitative);
- risk assessment and mitigation;
- documentation; and
- packaging and presentation.

As seen from the aforementioned list, technology is a limited piece in the puzzle. Program planning is not limited to technological skills alone; rather, it is a combination of organizational and logistical disciplines, business, technology, and human factors. Although there are general approaches to planning, if one considers the aforementioned variables there is no single solution to successful program development.

7.1 Executive Support

Executive support is a critical component not just to complete the strategic IT plan, but also for its implementation. Ongoing executive support is needed to advance and advertise the message and objectives related to the recommendations that have been selected for action. In fact, the ability to procure the resources (staff, time, and money) needed to execute the plan, and to remove or minimize obstacles along the way will require executive support. Without support, implementation of any significant actions will not be able to achieve the desired results.

Executive support is one of the first things a program manager should procure. The executive steering committee should be clear on their support for the plan's implementation, and if there is lack of support the reasons should be clearly understood and documented. Such support will aid in further development of the plan as needed resources will be easier to secure. Early executive support may result in scope modifications or refinement to better meet executive expectations and objectives; however, these changes will also result in some level of ownership and commitment. Executive support should also provide an ally for the program manager to turn to.

7.2 User Involvement

Second to executive support, user involvement is a significant target to engage. Users are the eventual true owners of a system and need to help define not only existing obstacles and opportunities, but the vision of the future state as well. Their involvement early on in the process helps garner support for the program as well as commitment of critical user resources in the plan's development, and helps to shape and prioritize activities and deliverables. In effect, a program manager is a facilitator whose job is to deliver a user's needs.

User involvement assists in defining and clarifying these needs (i.e., requirements) and focusing the associated scope and deliverables. Early on in the process, it is important to identify and engage the user base, or at least a representative group of them, to be an active part of the program process. If possible, it is important to make sure these users will be part of the program team through final acceptance of the system. A few key personnel should be selected to be the final "owners" of the system; these personnel should be made aware of this role from the beginning as they will be the "go-to" people when the program is closed.

7.3 Objectives and Requirements

For each recommendation, the objectives, goals, or changes must be clearly understood by all involved. Additional stakeholders who were not involved in the plan development will need to be briefed to get their input and support.

Moving from high-level recommendations to more detailed requirements will require additional efforts to solicit input from staff who will be users of the new tool or whose daily work processes will be affected by a new IT tool. Requirements will form the basis of organizational, process, and technology changes, and will dictate the scope, schedule, and budget of the program. Lack of clear requirements is a leading cause of project failure. Conversely, proper requirements are a leading factor in project success.

7.4 Strong Program and Project Management

Successful program and project management require skill and experience, not just within the technical realms of the effort but in the arts of resource management, scheduling, consensus building, problem resolution, contingency planning, and a myriad of other skills.

A common practice is to place the local technical expert in charge of a program or project because they are more familiar with the intricacies of the technology. Although this knowledge is needed, experience has shown that roughly one or two out of five technical experts are successful in the realm of project management on their first engagement. This is neither intended as a criticism nor to say that technical experts cannot perform as project managers, rather, that they must have the tools, training, and desire for this different role. The more critical the program or project, the more experienced and seasoned a selected project manager should be. A program and/or project manager is ultimately responsible for delivering a business improvement within the time and money allocated. The decisions and actions needed are often not technical in nature, but managerial and leadership based.

7.5 Scope Control

Scope control can be simply defined as working on those things needed to implement the IT plan. "Scope creep," or the gradual, often undetected and unfunded addition of functionality and desires, will likely affect a program near the end of the schedule when it is too late to recover without significant action being taken. Scope control has many facets, including

- clearly stating priorities up front (what will be done and what will not be done);
- keeping the scope as minimal as possible;
- breaking efforts and projects up into smaller, more manageable deliverables;
- keeping close tabs on work efforts and curtailing any expenditures that are not directly related to deliverables; and
- identifying where new scope is actually needed and formalizing appropriate actions to secure related changes in schedule, budget, resources, and expectations.

Scope control is critical to program planning, both as an up-front tool for appropriately communicating plans and resource needs and an execution monitoring and control tool to help ensure program success.

7.6 Drivers and Constraints

Drivers are the motivations or pressures that cause one to do something; drivers are typically the cause of one's objectives. A utility will typically not implement a program plan unless there is both a strong motivation and a justification. Such motivations, or drivers, can come from either inside or outside of the organization. Internal drivers could include such problems as the inability to manage massive amounts of data, lack of user interaction, or the inability to collect key metrics and make associated business decisions. External drivers could be public pressure to track operation and maintenance costs, the need to meet regulatory requirements, or customer dissatisfaction with publicly advertised information. As a first step in implementation, it is important to clearly understand the drivers involved in instigating the recommendation to support communications and measuring of results.

Constraints are factors that either limit what is done or cause it to be done in another way. Constraints can be either internal or external; in either case, they typically make a job more complicated. Internal constraints include staff skills and availability, lack of technology infrastructure, or limited funding or time. External constraints may include political or regulatory deadlines and commitments or public relations and reporting challenges. It is important to be clear about existing constraints before initiating implementation efforts. It is possible that the additional work and hurdles that must be overcome can reduce the ultimate value to the utility of the effort; it may be prudent to reprioritize efforts and modify the implementation schedule until the constraints can be reduced or removed.

It is important to capture and understand both drivers and constraints as they will shape implementation and are the ultimate basis for program expectations.

7.7 Change Expectations

A plan implies that something will be done or some change will be implemented. A plan is a description of the steps and resources involved in getting from one condition or state to another. To determine the steps needed to get from here to there, the definitions of the *current state* (i.e., as is) and the *future state* (i.e., to be) need to be completely and accurately understood. The program plan is then, effectively, the method and actions derived to get from a current state to a newer one.

Although understanding specific program details is important, there needs to be a reasonable limit to how much detail must be collected before action is taken. Indeed, “analysis paralysis” can lead to setbacks because of a perceived lack of progress. This is part of the art of developing a program: knowing when there is enough critical mass of information to take action.

To what level of understanding then should as-is and to-be states be understood? Whereas the initial level of understanding in the strategic planning process will be conceptual, for implementation it will be necessary to understand the business processes and data at a more detailed level. What needs to change can then be defined. The initial *gap analysis* is a useful starting point. Once what needs to change and why has been defined, the program manager can then define how, where, when, and who will change it. These elements represent the basis of a fully prepared program plan.

7.8 Business Environment

Whether the plan is to install a new supervisory control and data acquisition (SCADA) system or a new accounting process, it is part of the business of the utility and, thus, part of the utility business model. Because the program plan will have something to do with the utility business, it is important to keep the business context for the IT activities at the forefront and to refrain from using IT terminology when communicating with utility staff. If there are critical IT terms that are important for the utility to add to their vocabulary, this should be addressed with managers and human resources staff to allow time for staff to develop the familiarity needed to support the changes.

The plan development may include some high-level business process documentation to illuminate the context for IT needs; however, more in-depth process modeling can be valuable in support of the implementation of the plan. Modeling helps capture the essence of problems and solutions and helps to standardize communications for all parties involved. Ultimately, this is a form of communication and is needed to firm up the essence of the problem/opportunity of the as-is state and the definition of the to-be state, all of which are the basis of eventual time, budget, and resource needs.

Figure 4.5 shows a model of the basic functions involved in treating water, for example, which does not provide sufficient detail for understanding the role of IT. This modeling should be expanded and elaborated on until a complete understanding of the current state is understood as well as specific changes that will be implemented to improve outcomes. An example of the more detailed "swim-lane" model is shown in Figure 4.6. As these models continue to evolve, typically during program execution, it will be found that there are more sophisticated, automated, and interactive modeling tools that are available to, and used by, software programming staff. WEF's WISE program provides an explanation of the methods for modeling business processes based on utility experiences.

7.9 Stakeholder Engagement

In all aspects of program planning, three fundamental groups of people will ensure either the success or failure of the program plan. These are management, customers (including employees, who are often considered the customers of software projects), and service providers. In general, both the program plan and program manager must engage, align, and maintain all three groups during both the development and execution of the plan. The best approach to working with stakeholders of any type is to maintain regular and open communications. Honesty, transparency, and empathy are the three factors necessary to ensure that the plan achieves its goals. The

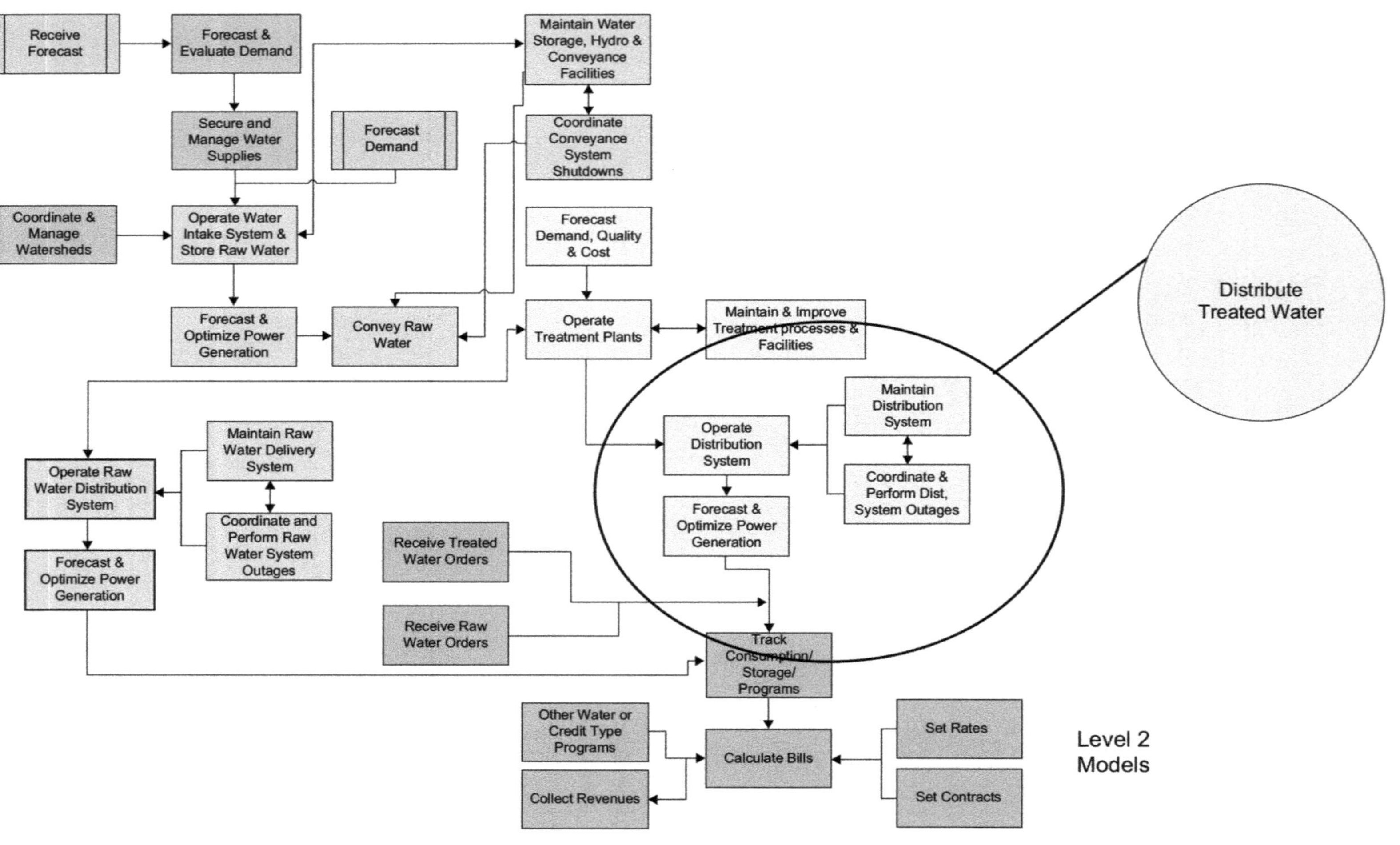

FIGURE 4.5 Sample Core Business Function Model (Courtesy of D. Henry, PE, MWD of So. CA, in concert with EMA Engineering)

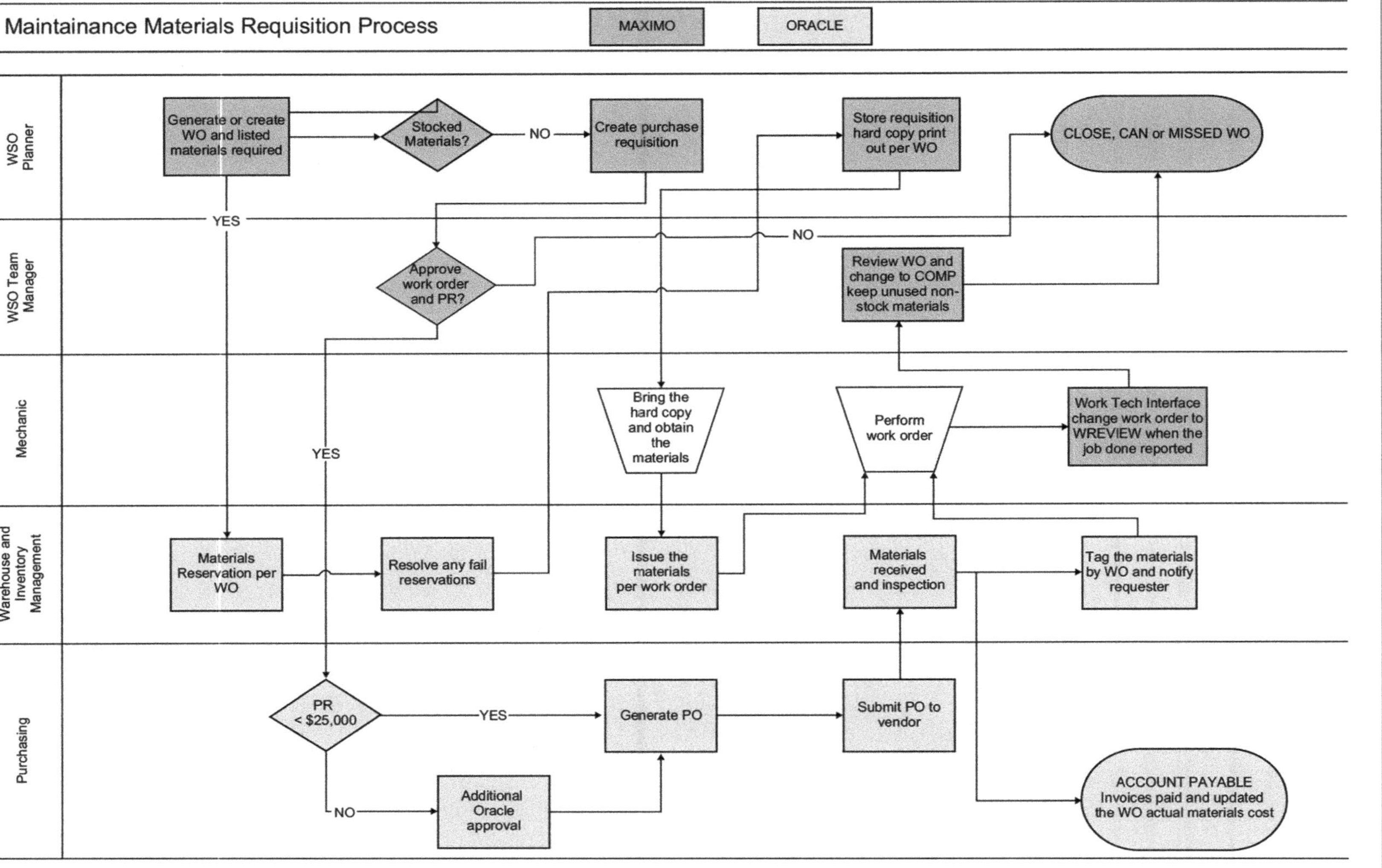

FIGURE 4.6 Sample Swim-Lane Model for a Requisition Process (Courtesy of S. Ma and D. Henry, PE, MWD of So. CA)

efforts to do business process mapping can be helpful in helping different groups understand and appreciate the need for change and creating a sense of teamwork. The earlier stakeholders are brought into the discussion of needs and priorities, the more likely they are to develop the perspective necessary to support the efforts.

7.10 Navigating Change

Once a plan is finalized and specific activities are set for implementation, there is often an increase in apprehension in those that will be affected. As the changes become imminent, they feel more real; the excitement of the plan's potential can be overshadowed by the fear of uncertainty. Most of the previously mentioned factors help in navigating this, and there are well-established processes for change management including Kotter's Leading Change and Prosci's ADKAR model. In addition, there are methods for understanding individuals to better tailor support activities, including the Myers-Briggs Type Indicator and Marston's DiSC model (1928). Figure 4.7 presents the elements of the Myers-Briggs personality classifications and their influence on behavior that will influence project dynamics.

It is useful to explore the different options and find the one that best fits the organization and the nature of the changes. Projects that include significant process changes in multiple departments in addition to new digital tools will require a much more extensive set of support activities than a project that affects a small subset of staff or that involves a change to IT infrastructure that users do not experience.

It is important to consider the changes and proactively estimate the time and attention necessary to support to achieve the desired outcomes. Even if people do not visibly resist change, acknowledging the human factors of the project can bring dramatic improvements in outcomes. The plan is not just about document actions and outcomes; it is about getting everyone moving in the same direction toward a new state of business and arriving there successfully. The plan merely provides the agreed road map, deliverables, and timeline that can be referred to during execution.

In defining that road map and timeline, it is important to include activities that address the effect of the changes on all stakeholders. This could be meetings, focus groups, or communications. The common elements across the various models include the following:

- Make sure everyone affected understands what the change is, when it will happen, and what the expectations are from them individually and as a group.

Spectrum of Type Preferences
& Area of Influence on Behavior

Extraversion (E) Obtain energy from people, activities, and things outside of oneself. Share energy easily with others. Prefer to talk things though.	Source of Energy 	Introversion (I) Gain energy, ideas, and inspiration from within oneself. Can be drained by too much external stimulation. Prefer to think things through.
Sensing (S) Absorb information through the 5 senses and prefer tangible facts.	Processing Information 	Intuition (N) Big picture oriented. Process external signals as possibilities and notice patterns and relationships.
Thinking (T) Approach decisions objectively based on logic. Focus on the issue at hand and detached from the emotion of decisions.	Decision Making Style 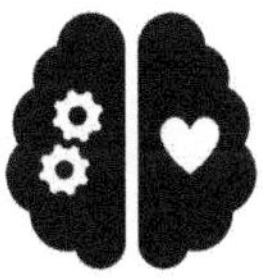	Feeling (F) Approach decisions with empathy. Consider impact on the people involved. Look to understand and avoid conflict.
Judging (J) Prefer order and closure on decisions. Not comfortable with open ended outcomes.	Perspective on External World 	Perceiving (P) Maintain flexibility in interacting with the external world. Are adaptable and enjoy generating many possibilities.

FIGURE 4.7 Myers-Briggs Type Indicator Elements (Adapted from Tucker, 2008)

- Provide context for the reason for the change and the benefit to the individual to support the change.
- Provide the support (i.e., training) for stakeholders to be able to adapt to the change.
- Maintain support and dialog until the change sticks.

Adequate time and resources need to be given in the implementation for these activities, as well as buffer time to address any roadblocks that may arise. Although it can be tempting to ignore input from a stakeholder who resists change, there are often elements of useful insight that can improve upon the project approach with proper consideration.

As important as the methods of change management are the nature and tone of the project leaders. Here are some important considerations in selecting and preparing someone to lead the implementation of the plan:

- How extensive will the effects be on staff and other stakeholders and over how long a period of time? For long-duration projects and those with significant change expectations, it can be more effective to dedicate an individual with experience navigating change to support these activities so the plan project manager can focus on other aspects of the implementation.
- Is the person empathetic and effective at hearing feedback? The activities necessary to support change are not punch list items. They require attention and human interaction, which can be messy and unpredictable. Even when an area is not open to debate, there are people who want to debate. Simply hearing them and acknowledging them can go a long way.
- What communications skills are necessary? Communications are critical to success and need to be done frequently and in the modes that resonate best with each stakeholder group. People typically need to hear something a number of times before it can be processed. If there are communications professionals available to support the lead, this can be effective; otherwise, the leader of the implementation will need to have or learn basic change communications skills.
- Define the governance process for the project. Be clear about who has the ultimate final word on each component of the plan's implementation. If some things are not negotiable, such as having to upgrade a system or putting cybersecurity protocols in place, be clear about that, and highlight when there is an area where stakeholder input can guide decision making.
- Trust, honesty, and transparency are critical to success. Taking time to establish trust among the project team is time well spent. If there are unknowns, be clear about them, and do not hide information from the team because that deteriorates trust. If there are areas where information must be kept private for privacy or security reasons, let people know; they are likely to respect that.
- Do not strive for perfection. Each individual's perspective on perfection is different, and perfection is expensive and should be reserved only for high-stakes (i.e., life-or-death) activities. Continual improvement is the goal, and creating a learning environment that supports

staff through the never-ending changes of running a utility is the best investment of time and energy.

At the organizational level, change is typically a question of logistics and how much an organization can accommodate change while still performing its day-to-day functions. It is typically a question of how many resources the organization can spare to effect the change. If there are not enough resources, either more resources need to be brought in or the change needs to be throttled back to an acceptable pace. This condition is not necessarily limited to technical resources; it may be that the availability of legal contracts, purchasing, operational, and customer resources are as likely to cause delays as technical efforts. This effect is typically a function of project life cycle, timing, and magnitude, and depends on what stage the program is in and where the resource bottlenecks are created. Therefore, a program manager should plan and communicate accordingly and manage scope and expectations, as discussed previously.

8.0 REFERENCES

Government Finance Officers Association. (2005). *Recommended budget practice on the establishment of strategic plans.* Retrieved August 2009 from http://www.gfoa.org/downloads/budgetStrategicPlanning.pdf

Marston, W. (1928). *DiSC: Dominance, influence, steadiness, conscientiousness.* Inscape Publishing.

Tucker, J. (2008). *Introduction to type and project management.* CPP, Inc.

Vitasovic, C., Olsson, G., & Haskins, S. (2021). *Water intrapreneurs for successful enterprises (WISE): A vision for water utilities.* Water Environment Federation, Leaders Innovation Forum for Technology, & Water Research Foundation.

9.0 SUGGESTED READINGS

Brueck, T., Rettie, M., & Rousso, M. (1997). *The utility business architecture: Designing for change* (Project No. 165). Water Research Foundation.

Collins, J. C. (2001). *Good to great: Why some companies make the leap—and others don't.* HarperBusiness.

Gleicher, D., & Beckhard, R. (1969). *Organization development: Strategies and models.* Addison-Wesley.

Goodstein, L. D., Nolan, T. M., & Pfeiffer, J. W. (1993). *Applied strategic planning: A comprehensive guide: How to develop a plan that really works*. McGraw-Hill.

Standish Group International, Inc. (2001). *Extreme CHAOS*. Retrieved August 2009 from http://www.smallfootprint.com/Portals/0/Standish-GroupExtremeChaos2001.pdf

SWOT analysis. (n.d.) In *Wikipedia*. Retrieved August 2009 from http://en.wikipedia.org/wiki/SWOT_analysis

U.S. Environmental Protection Agency. (2017). *Effective utility management framework*. watereum.org

Vitasovic, Z. C., Horne, J., Kricun, A., & Haskins, S. (2018). *Management of water sector utilities: Summary of industry initiatives and research* [White paper]. Water Research Foundation.

Zachman, J. A. (2006). *Enterprise Architecture* home page. Retrieved December 2009 from http://www.zachmaninternational.com/index.php/home-article

5

Organizational Aspects of Information Technology

1.0 SUMMARY OF KEY THINGS TO KNOW

- Well-defined and well-understood roles and responsibilities enable the effective use of technology.
- Many information technology (IT) responsibilities are not assigned to IT staff. All utility staff have important IT-related roles.
- Organizational structure affects how technology is selected, managed, and supported.
- Clear governance for IT activities and decisions results in smoother operations.
- Any technical challenge can be overcome with a trusting and supportive culture.

2.0 ROLES AND RESPONSIBILITIES FOR INFORMATION TECHNOLOGY IN A UTILITY

There are many different models for IT responsibility that can involve utility staff, external resources shared with other local or regional government entities, and private business. It is not uncommon for relationships between IT and other utility staff to be strained. When there are misunderstandings and mismatched expectations, it is often because of the lack of clear assignment of responsibility for IT-related activities including decision making, business applications, and data management. This can lead to a disconnect in understanding the strategic importance of IT to support water and wastewater services and a lack of effectiveness in meeting the needs of the communities that they serve. This section will describe the key roles and responsibilities associated with well-managed IT programs.

2.1 Users

There are two categories of users of IT systems: those who enter the data, and those who use the outputs from the system. The users of each system vary based upon the business application. Users who input data into a system are responsible for providing the required data accurately and in a timely manner. They are also responsible for any other system functions that are part of their job. For example, an accounting staff person may upload or enter debits and credits into the financial system and be responsible for running reports and conducting analyses of running financial expenditures. Users who are consumers of data from a system are responsible for making decisions and taking action based upon their job requirements. An input user of one system could be a user of outputs from another. For example, the accounting staff mentioned above could be a consumer of reports generated by the computerized maintenance management system (CMMS) about the cost of work order completion. Some examples of responsibilities are presented in Figure 5.1.

In general, users' collective responsibilities include

- entering accurate and timely data,
- using the software as designed to complete business processes for which they are responsible,

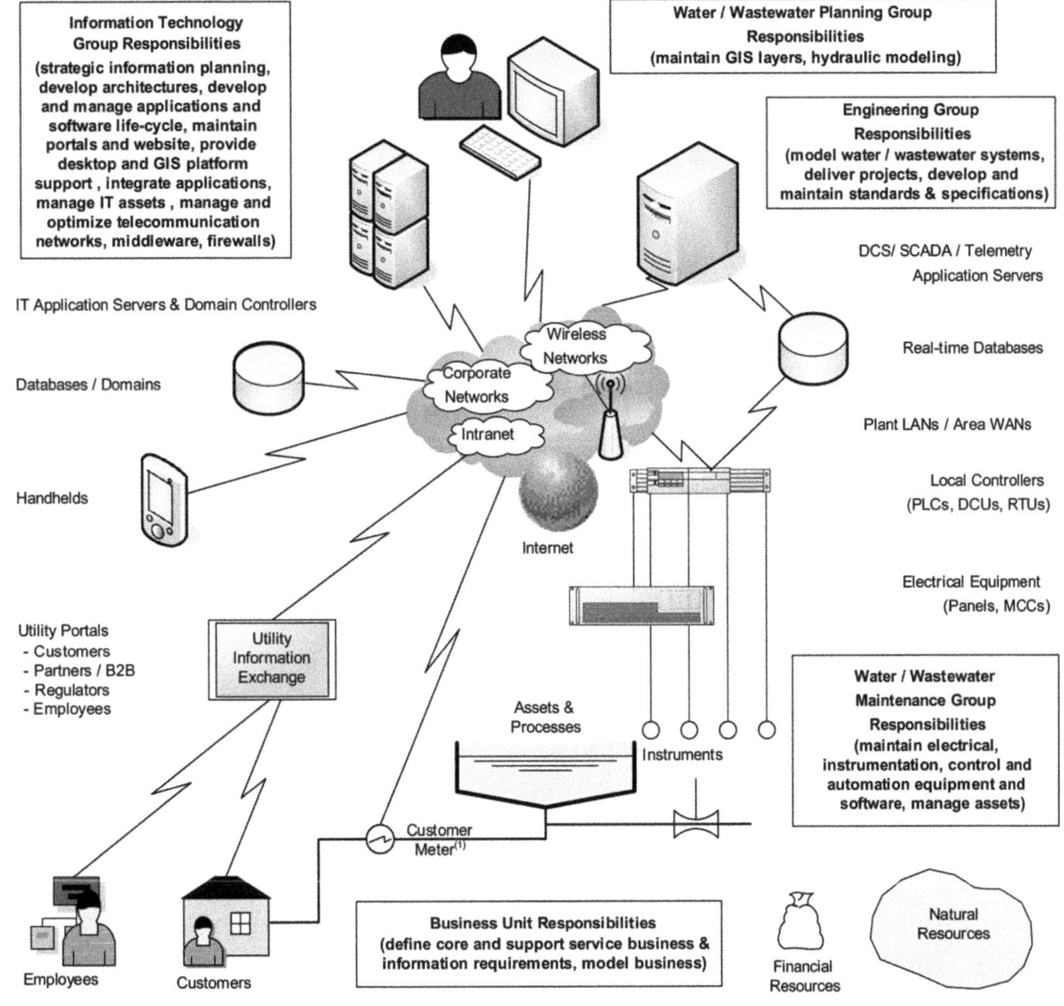

FIGURE 5.1 Responsibility Allocation for IT Services in Today's Typical Utility

- communicating with other users of the system if there are issues to resolve or questions to address,
- participating in evaluating and selecting new IT tools,
- participating in testing tools as upgrades occur,
- formally reporting problems to IT support staff, and
- understanding the processes and procedures necessary to effectively use digital tools to complete job requirements.

It can often be the case that there is shared responsibility for an IT application. An example is a CMMS that integrates data sets from multiple systems and whose users span multiple departments. In addition, in some cases a system like a CMMS may be shared across municipal departments as well, including public works, water and wastewater departments, and geographic information systems (GIS) staff—each with different needs as users of the system. Even within a single utility there can be diverse users including maintenance staff, planners, project managers, and engineers. Clearly specifying the roles and responsibilities of each type of user is critical to ensuring the effective use of the system.

2.2 Data Stewards

All system users have data stewardship responsibilities. It is also best practice to have a specific subset of users responsible for periodically evaluating data quality and reviewing the current and emerging uses of the data to validate the data as appropriate for its intended use. For example, GIS data are often collected using aerial photography for use to support fieldwork. It is the data stewards' responsibility to understand if the data are spatially accurate and timely enough for the work to be done. Primary responsibilities of data stewards include the following:

- documenting metadata. Metadata describe the data set with sufficient detail to inform appropriate use of the data.
- assessing and documenting the dimensions' data quality
- developing recommendations or taking action to correct problematic data
- communicating with users to maintain data quality

2.3 System Availability

Information technology systems are necessary to run a utility, so it is critical that the system is available whenever it is needed. This could be during normal business hours, or 24/7 depending on the role the system plays. There are a variety of different models for ensuring system availability as well. If an application is provided as software as a service (SaaS), the contract with the vendor supplying the system will specify the responsibilities for ensuring system availability and response times to issues.

Another model is that there is a centralized IT department for a municipality and resources are shared with the utility. This model can be a challenge for utilities because their priorities are in competition with those of other departments including police, fire, and public works. Best practice is to have open channels of communications, documentation of roles and responsibilities, and defined service-level agreements (SLAs) to ensure that utility operations are not disrupted. This will specify which issues utility staff can address, and which the IT department will address and within what time frame.

The third model is for the utility to have its own IT department and staff. These staff may interface with vendors for support and work with IT staff of the municipal government for specific systems or issues but have greater authority and control to manage the utility's systems and ensure system availability.

Regardless of the governance model, each system must have a designated contact to address system availability with the authority to act to bring systems online to minimize business disruption.

2.4 Hardware and Networks

The models for roles and responsibilities for acquiring, installing, and maintaining IT hardware and networks are similar to those described above regarding system availability. There are typically different staff responsible for IT systems than those responsible for operational technology such as supervisory control and data acquisition (SCADA) because of the difference in the technology and the training need to support them. Chapter 7 provides an in-depth discussion of the nature of hardware and networks in utility IT. There is not a single model of responsibility that would be considered standard practice; however, it is best practice to have staff designated with the responsibilities for IT hardware and software, either in-house, with shared resources, or with a private contractor.

Regardless of the governance model, the team that supports this will have responsibilities that include

- specifying hardware requirements necessary to run the user's software including servers, personal computers, and peripheral devices;
- supporting connectivity between the hardware, network, and internet to enable software to operate as intended;
- supporting any sensors that are necessary to gather data and integrate with software;

- monitoring network performance and addressing risks and poor performance events;
- maintaining a schedule for hardware maintenance and replacement and communicating the necessary budgets for this to management; and
- troubleshooting issues and getting service for IT equipment that is failing or underperforming.

2.5 Maintenance and Updates

All systems will require updates and system maintenance over time. This includes fixing bugs, issuing security patches, and releasing new functionality. Responsibility for these activities is shared across a number of different groups and roles:

- Software vendors are responsible for communicating the timing and nature of the changes.
- Information technology staff are responsible for assessing the effect on other systems and users, especially when systems and data sets are integrated across systems.
- Information technology staff are responsible for involving the appropriate staff in testing the changes and, if possible, for accepting and scheduling changes. Some changes are mandatory to protect system security or to maintain software service contract terms, but others may be optional.
- Users are responsible for testing changes and potentially taking training to learn how to work with the changes.

2.6 Selection of Digital Tools

The selection of digital tools requires a diverse team of stakeholders including IT staff, users, vendors, and senior management. The process for selecting digital tools is described elsewhere, but the roles and responsibilities of the team involved in the selection include the following:

- Identify the need, which may be an opportunity or gap, that the tool addresses for the business.
- Articulate and prioritize the requirements for the tool to meet those needs.
- Understand the business processes associated with the tools and evaluate how process changes will support the goals of the tool.

- Evaluate the tool options and participate in vendor demonstrations.
- Research how peer utilities are meeting the identified need.
- Obtain senior management support to understand the broader perspective of the need, and the cost to meet it, compared to other utility priorities.

3.0 ORGANIZATION AND GOVERNANCE MODELS

The organizational and governance challenges faced by utility IT departments can be significant. A plethora of hardware, software, databases, networks, and other IT systems have to respond to a set of complex business requirements that demand a myriad of resources and skills. Information technology services can range from calibrating instruments in a treatment process to making strategic decisions in the boardroom.

3.1 Organizational Considerations for Utility Information Technology

The top 10 organizational challenges facing a typical utility IT department today include (1) integrating the utility IT vision with the utility business vision; (2) lack of full appreciation of the strategic significance and potential of IT because of "engineering orientation" of utility management; (3) lack of a strong IT governance framework to guide IT investments and ensure adherence to architectures and policies; (4) lack of an IT "voice" on the utility's executive board; (5) insufficient IT resources that truly understand the utility's operational and business requirements; (6) technologies that are changing faster than customer organizations can adopt them; (7) a significant technology awareness and usage gap between the utility's senior management and the latest generation of technology-savvy employees; (8) lower IT pay levels in public-sector utility organizations than in the rest of the IT industry; (9) attracting and retaining qualified IT resources; and (10) the need for IT staff to work well with varying cultures of employees from planning, engineering, operations, maintenance, and finance.

Other related challenges include cybersecurity, IT budget pressure, the politics of offshoring (especially in the public sector), use of SaaS and other outsourcing, and legislation. Local factors determine the priority level of each challenge.

In many utilities, the roots of IT challenges can be traced back to more than just organization and governance. Restrictive procurement policies, civil service constraints, and union issues have also been significant hurdles to realizing the full strategic potential of utility IT organizations.

3.1.1 Municipal Utilities

Information technology challenges are typically greatest for water and wastewater departments governed within larger municipalities that deliver a broad range of public services. Those departments often compete for IT resources, funding, and priorities. In addition, they sometimes need to create and maintain more working partnerships with other front-line service organizations to reflect their own needs in shared applications and data. Examples of this include sharing a CMMS with the roads and highway department, an enterprise asset management system with the finance department, or a GIS with various organizations in public works and corporate services.

A leading practice is to adopt a customer-focused "IT associate" or customer-relationship manager model, in which IT resources are co-located with front-line water and wastewater operating groups while reporting into corporate IT departments. Another leading practice is for the utility to have its own IT staff, who work as part of a broader municipal IT team but who are dedicated to utility needs and support.

3.1.2 National Utilities

There are a number of countries where water and wastewater needs are met by a single nationwide utility. These national water and wastewater utilities are often dependent on various government ministries for resources, policy direction and data, and application support. In developing countries, these utilities often have difficulty finding local resources to meet their IT needs. For instance, when these utilities look for solutions by hiring vendors, consultants, and contractors to cover the shortfall, temporary rather than sustainable relief is provided.

At one large national utility in the Caribbean, for example, consultants and contractors have installed a large variety of instrumentation and control equipment during the past 20 years. When this is done, however, it is critical that consideration be given to the maintainability and the local capacity to service the equipment. The same can be said for a number of IT applications and technologies at this utility. Utilities in other developing countries often face similar challenges. Access to SaaS models can alleviate some of this.

3.1.3 Publicly Owned and Privately Operated Utilities in North America

Private companies operating and maintaining utility assets for customer organizations may be dependent on that client's IT resources for data and access to external IT systems and services. Sometimes, this occurs in an environment in which responsibilities for providing and maintaining data, managing databases, and maintaining IT systems are not well documented in operating agreements. They may also have to provide IT resources with the skills to manage a combination of customer-mandated systems, such as CMMS and SCADA, and their own preferred systems. These preferred systems are sometimes used as standard tools by private operators to ensure consistent and efficient service delivery for customers across a region or nation.

3.1.4 Privately Owned and Operated Utilities

Fully privatized utilities have complete responsibility for all aspects of IT required to efficiently and effectively deliver water and wastewater services. Their approach to IT has changed considerably since privatization in 1989 and has also varied greatly between utilities.

During the first postprivatization decade, most of the 10 large private water utilities grew their internal IT organizations and service offerings beyond anything traditional utilities had ever seen. One went as far as developing its own remote terminal unit hardware, whereas others created and spun off separate IT organizations that provided services and specialized application software to the originating utility, other UK utilities, and utilities in other countries.

This expansion and divestiture of IT caused special challenges for core utility organizations. Should they acquire and use only their own proprietary IT resources and software solutions, or could they use other vendors? What IT role would remain in the utility? And how would their service requests be prioritized by a more arm's-length IT service provider?

During the second postprivatization decade, a number of large water utilities narrowed the scope of IT products and services they delivered, the business sectors in which they were offered, and the geographic markets in which they serviced. Thames Water, one of the world's largest water and wastewater utilities that serves customers in and around London, England, for example, returned to a focus on delivering core water and wastewater services in its domestic UK market, supported by internal services from its IT department. For utilities like Thames Water, organizational challenges in IT are moving back to the more traditional utilities mentioned previously. Others retain their external IT service delivery companies but focus their efforts on serving the water industry.

3.2 Governance

A typical utility today has a large number of individuals and groups with IT-related responsibilities as described in Section 2. In addition to clearly defining and communicating roles and responsibilities, it is best practice to have processes to govern IT decision making. Areas where governance should be defined include the following:

- New technology investments—Requests for new technology can be initiated from anywhere in the organization. A process should exist to direct requests to a central person or team for evaluation, and to triage requests based on the extent and magnitude of the investment. Information technology must have a role in that evaluation to ensure compatibility with existing systems and ensure security, and someone with budgetary authority should also be included to assess the benefit of the investment in comparison to other needs. A process that has a filter to route requests based on complexity and cost is advisable. It is important that the process not be overly cumbersome so the cost of the governance process does not exceed the cost of the request.
- System access decisions—As new users join the organization or if requests for system and data access expands beyond the initial user base, a process is needed to determine who can have access to view and interact with the systems. A robust governance process will include that system's business owner, who would review the request and consult with appropriate staff. Then new users could be added by the designated system administrator, who could be a power user, an IT staff person, or a software vendor representative.
- System integration decisions—There is great potential for IT to add business value for a utility by integrating data sets and system capabilities across software platforms. System integrations, however, require more than just IT programming. They often require modifications to business processes, user responsibilities, and the addition of new analyses and processes. When systems are integrated, it is often the case that staff and units that operated independently are now interdependent. There should be a governance process to evaluate the effects of system integrations and assign responsibilities for implementing the non-IT-related changes.
- Use of social media—Social media is used by most utilities to communicate with customers, and many staff are active on social media platforms. There should be a policy and governance about what is communicated on social media and who is authorized to post on behalf of the utility. Improper use of social media can cause public relations issues and should be addressed in a systematic manner.

There are also many governance considerations related to cybersecurity, which are discussed in Chapter 8.

A clear and complete IT governance model enables a utility to proactively meet challenges that arise and appropriately allocate IT responsibilities using a strategic and integrated organizational response. Information technology's organizational maturity in terms of culture, management style, customer relationships, communication, partnerships, and organizational structure and IT governance has a major role to play in meeting this response.

4.0 ORGANIZATIONAL MATURITY MODEL

The eight-step process described in this section provides a systematic way for utility IT organizations to understand how effective their organization is with respect to IT, and to undertake a successful transformation from the current "as-is" to the target "to-be" maturity level, organizational structure, and governance framework. It also helps those organizations identify and apply ways to respond to opportunities and challenges. Each of the following steps can be focused and adapted for specific utilities and situations:

1. Define and characterize the current state.
2. Identify, discuss, and inculcate the future state.
3. Develop strategies that leverage workforce and technical trends.
4. Select your organizational structure.
5. Fill new positions and build better connections.
6. Get strategic and raise the profile.
7. Decide who does what.
8. Manage the change.

4.1 Define and Characterize the Current State

An industry-accepted way of defining and categorizing the state of an IT organization is the capability maturity model (CMM). In this manual, the CMM has been adapted to focus on organizational characteristics rather than including all the service-related characteristics. Table 5.1 shows the organizational maturity matrix (OMM), which describes the organizational characteristics of utility IT organizations at various levels of maturity.

To determine how close to fully "mature" an IT organization is, the following questions need to be asked in the context of "How close are we to having a . . .":

- Collective sense of IT mission, strategic IT future vision, both with a dynamic relationship to the utility's mission and vision?

TABLE 5.1 Organizational Maturity Matrix for Water and Wastewater Utility IT (Adapted from Capability Maturity Model)

Organizational Maturity	Description	Organizational Performance Indicators
1 IT tasks	• Reactive work • Task focus • Results dependent on individuals • Least productive	No collective sense of IT mission or IT future vision Culture unidentifiable, widely variable Structure loose, decentralized, decisions based on influence of individuals Management role unclear, inconsistent, may contribute to projects, daily IT tasks Communication with staff and employees inconsistent and infrequent Customer relationships based on connections between individuals, customers seen as passive recipients of IT services, customer experience varies greatly
2 IT projects	• Some routine work • Project focus • Some standard results, but variable between groups • Results delivered by project teams and work groups	Little collective sense of IT mission, typically no strategic IT future vision Culture hard to identify, strong and competing subcultures may exist Structure hierarchical, decisions made based on position power Management more removed from projects and daily IT functions Communication with IT staff infrequent, often only when something is wrong Customer relationships based on connections between groups and individuals, customers seen as recipients of IT services and participants in IT projects, customer experience varies by project and work group

3 IT business	• Proactive, planned, and systematic work • Program and service delivery focus • Consistent, standard results • Results delivered by everyone in IT following methods and standards on projects and other services	Collective sense of IT mission, typically no strategic IT future vision Culture likely strong, well defined, and shaped by standards and procedures Structure hierarchical and bureaucratic Management remote from daily functions of the organization Communication with IT staff done through a series of formal directives and regular staff meetings or presentations Customer relationships based on standard expectations built over time, customers seen as recipients of IT services, some periodic feedback received, customer experience fairly consistent
4 Utility business IT	• Planned work driven by customer needs • Utility focus, IT supports efficient business processes • Monitored and managed performance of IT services • Results delivered by everyone in IT focusing on customer relationships and effect on business performance	Collective sense of IT mission, strategic IT vision, reflects utility's mission, vision Culture strong, tightly defined, shaped by commitment to excellence, some ability to change Structure flat, flexible, customer oriented, some ability to change Management involved in daily IT functions only as facilitator and coach Communication with IT staff continuous and informal, with periodic directives and staff meetings to augment daily communications Customer relationship based on working partnerships with joint translation of business needs into IT solutions. Feedback is integral part of the relationship, customer experience well managed through surveys, periodic joint sessions Strategic leadership role of IT in utility beginning to take shape through organizational visibility of IT (e.g., direct report to top executive), strategic IT roles (e.g., CIO)

(continued)

TABLE 5.1 Organizational Maturity Matrix for Water and Wastewater Utility IT (Adapted from Capability Maturity Model) (*Continued*).

Organizational Maturity	Description	Organizational Performance Indicators
5 Utility strategy IT	• Work driven by innovation, anticipated change, and strategic customer needs • Utility focus, IT introduces new, innovative IT processes and initiatives • Results delivered by everyone in IT focusing on customer relationships, business performance improvement, and jointly inventing the utility of the future	Collective sense of IT mission, strategic IT future vision, dynamic relationship to utility's mission and vision Culture strong, tightly defined, consistent with corporate culture and values, shaped by commitment to innovation, creativity and excellence, change comes naturally "Soft" structure, networked, customer oriented, flexibly integrated with rest of utility IT leaders help formulate and sponsor new initiatives, reinventing IT, integrating IT into strategic business direction and leading change Communication with staff and employees is continuous and informal with periodic directives and staff meetings to augment daily communications Customer relationship characterized by close working partnerships in all stages of the IT and business solution life cycle. Bidirectional, dynamic feedback is an integral part of the relationship, customer experience continues to improve Strategic leadership role of IT in utility visible, enabled through strategic governance structures that include all utility business leaders, strategic IT roles (e.g., CIO)

- Strong culture, tightly defined, aligned with the corporate culture and values, shaped by a commitment to innovation, creativity, and excellence, to which change comes naturally?
- Structure that is "soft," networked, customer oriented, and flexibly integrated with the rest of the utility?
- Leadership team in IT that helps formulate and sponsor new initiatives, regularly reinvent IT, and integrate IT into the utility's strategic direction and lead change?
- Communication with staff and employees that is continuous and informal with periodic formal forums and staff meetings to augment daily communications?
- Relationships with customers characterized by close working partnerships at all stages of the IT and business solution life cycle, using multidirectional, dynamic feedback as an integral part of the relationship while each customer's experience continues to improve?
- Visible strategic leadership role for IT in the business of the utility, enabled through strategic governance structures that include all utility business leaders, recognizable by the presence of strategic IT roles (e.g., chief information officer [CIO] and vice president of IT)?

Although these questions are all in organizational areas that are difficult to measure, the areas should at the very least be gauged by using a simple, repeatable survey based on the OMM shown in Table 5.1. The process of asking questions and having discussions about survey results provides valuable contributions to positive change. Use of more quantifiable metrics to track outcomes such as productivity increases and improvements in service levels could be applied at utility-wide and team-specific levels. Additional metrics could also be developed as specific opportunities are identified and tracked.

The maturity level of IT organizations in many utilities is low when measured against the model shown in the OMM. It shows organizational performance indicators for each of five levels of maturity, including

- Level 1, which focuses on performing IT tasks;
- Level 2, which focuses on delivering IT projects;
- Level 3, which focuses on delivering IT services to deliver on IT's mission and vision;
- Level 4, which focuses on applying IT to deliver water and wastewater services to deliver the utility's mission; and
- Level 5, which focuses on integrating IT as a fundamental strategic element of the utility's future vision.

Many utility IT organizations find themselves at Levels 1 or 2, some at Level 3, even fewer at Level 4, and very few at Level 5. Therefore, for most of these organizations, unifying the entire IT response to utility business needs under an umbrella of shared strategic direction and unimpeded teamwork would provide significant benefits. Organizationally, that would require raising the corporate profile of IT, better defining IT roles and responsibilities, and improving the partnerships between corporate IT resources, IT resources in front-line departments like water and wastewater, and the various functional specialists throughout the utility. It would also require creation of an inextricable link between the IT organization and its services to the delivery of the utility's mission and strategic vision.

4.2 Identify, Discuss, and Inculcate the Future State

The utility of the future would deliver IT through an organization that has adopted all the characteristics of the mature Level 5 organization, as described in the OMM and the traditional CMM. It would also anticipate trends in workforce and technology, leverage opportunities provided by the marketplace, and consider strategic implications of global trends on water and wastewater.

The IT organization's vision would take all those elements into account and be fully aligned with the overall utility's vision. Table 5.2 shows the municipal IT vision and mission for the city of Palm Coast, Florida. The city is focused on the contributions IT will make to front-line department-service outputs and program outcomes, innovative contributions to those outcomes, and a commitment to delivering enterprise-wide and leading IT solutions.

For IT departments in nonmunicipal utilities, the vision and mission could be even more specific, referring to residential, industrial, commercial, and institutional customers; regulatory reporting; and the use of the utility portal to build on customer relationships through ready access to information.

The principle of involvement is important to follow when creating a vision and mission for the IT organization. If possible, everyone in the IT organization should be involved while keeping the customer in mind. In addition, customers should be involved for input and feedback. Indeed, the more mature utility organization will increase involvement of its customers in strategic planning.

4.3 Develop Strategies That Leverage Workforce and Technical Trends

It is important for utility IT organizations to leverage workforce and technology trends when developing strategies to reach their vision.

TABLE 5.2 City of Palm Coast, Florida, IT Vision and Mission Aligned With Utility (Department) Vision and Mission

Water and Wastewater Utility	Corporate Information Technology
Vision	Vision
As the community grows, so will the utility. The utility department is composed of a highly trained professional staff that is prepared to provide the best level of service to the community today and into the future. While using the latest technology, along with proven industry standards, we will continue to develop the infrastructure to meet the needs of the growing community.	The information technology and communications department will be a proactive leader, identifying issues and offering innovative solutions to enable city departments to accomplish their goals and provide quality services to our citizens more effectively and efficiently.
Mission	**Mission**
Our mission is to provide safe drinking water and the best wastewater service to our customers at the lowest possible cost while adhering to the strictest guidelines for water quality and environmental protection.	Information technology is committed to serving the business operations of the city by providing enterprise-wide, integrated solutions with emphasis on superior customer service. Ensure effective and efficient use of new and existing technology resources and investments. Exceed internal and external service expectations by implementing leading-edge solutions in line with established "EGov" best practices.

4.3.1 Widening Generation Gap

Water and wastewater utilities have known for more than a decade that there will be a significant shift in the demographics of staff. The differences in expectations and fluency with digital tools varies dramatically between tenured staff who are approaching retirement and the new generations of workers who will replace them. With the rapid pace of technology, this gap is not likely to go away with retired staff; it will only persist in different forms with different expectations and standards. The COVID-19 pandemic resulted in a dramatic shift in the acceptance and adoption of IT tools for staff of all experience levels and ages to keep the utilities running while offices were closed. However, comfort levels with new technologies will continue to evolve.

4.3.2 Ubiquitous Technology

The increasingly ubiquitous nature of IT in the world and workplace will affect utility IT organizations significantly. New employees will expect to be fully supported by the latest technology and the best decision-making tools.

Otherwise, they may choose to work at more technology-savvy organizations. Operators and maintenance workers will expect the availability of process conditions, equipment, and recent laboratory test results on their handhelds as they're doing their rounds. In addition, network technicians will expect to be able to remotely diagnose and even repair any potential problems in corporate and local networks.

Front-line staff will also be able to contribute more of their own skills in configuring technology tools as knowledgeable self-supporting users. This can, in turn, contribute to decentralization of a number of IT services.

4.3.3 Changing Workforce

Strategic IT organizations need an understanding of other global trends in the workforce as well including the following: the reduced availability of technical IT staff, both locally and globally; the changing aspirations and attitudes of the IT workforce (Salkowitz, 2008) as a new generation of staff is hired; the changing ethnic mix in the workplace; and the role of outsourcing and "offshoring" of IT services and the shift to virtualization of computer systems in which support staff may be located in a different time zone or country. A benefit of workforce trends is that they provide an opportunity for a utility to have a team of staff who are well aligned and representative of the diverse communities they serve.

In the best utilities, IT leaders provide strategic advice on how the business might anticipate and apply advanced technologies, from self-cleaning remotely diagnosed dissolved oxygen probes to embedded state-of-the-asset reporting technologies to keep the business at the forefront of providing value to customers and shareholders.

4.4 Select Your Organizational Structure

There are three recognizable design models for utility IT organizations to consider adopting for their journey to reach the vision. These are

- the *department model*, where most IT services are decentralized and corporate IT is responsible for the desktop, the corporate network, and the applications and IT services not covered by front-line departments. From an application perspective, corporate IT is only responsible for those applications for which distributed departments cannot provide support and have requested assistance. Major applications are the responsibility of front-line departments. For example, financial management systems are the responsibility of the finance department, the operations department is responsible for the operations management system, the laboratory is responsible for the laboratory information management system (LIMS), and operations and maintenance is responsible for CMMS. Most IT financial and human resources are allocated to the line departments rather than corporate IT.

- the *shared model*, where IT services are shared between corporate IT and front-line departments and where corporate IT is responsible for the desktop, the corporate network, and a series of agreed-upon enterprise-wide applications and IT services. From an application perspective, corporate IT is responsible for key enterprise-wide applications that are determined to be better managed for the good of the overall utility. Typically, this would include systems such as financial systems, human resource systems, GIS, and enterprise asset management systems. Specialized applications are the responsibility of front-line departments. For example, the operations department is responsible for the operations management system and the laboratory is responsible for the LIMS. Information technology financial and human resources are allocated evenly to the line departments and corporate IT.
- the *enterprise-wide model*, where corporate IT is responsible for creating and managing an enterprise-wide information architecture; setting and enforcing enterprise-wide standards for integration, telecommunication, security, data, and information management; and the provision of all IT services. From an application perspective, corporate IT is responsible for all major applications, including operations management systems, LIMS, and CMMS. From an information perspective, it would take data from SCADA and process control systems, likely handed off in time-based packets, for it to be used in all other systems on the integrated network. The remainder of the utility organization would be responsible for defining all their respective departments' business requirements. Most IT financial and human resources are allocated to corporate IT rather than the line departments.

Figure 5.2 shows the aforementioned three organization design options mapped against the OMM. Which option is best is dependent on a number of design factors specific to each utility and the capacity of the IT department and the IT service providers available in the marketplace.

In addition to these three distinct models, there are an infinite number of combinations that can be used. Hybrid models would result from application of the more detailed IT decision-making matrix (i.e., establishing a contract/contractor management hub).

As the organization design process moves closer to the preferred high-level structure, decisions also need to be made at the next level of the

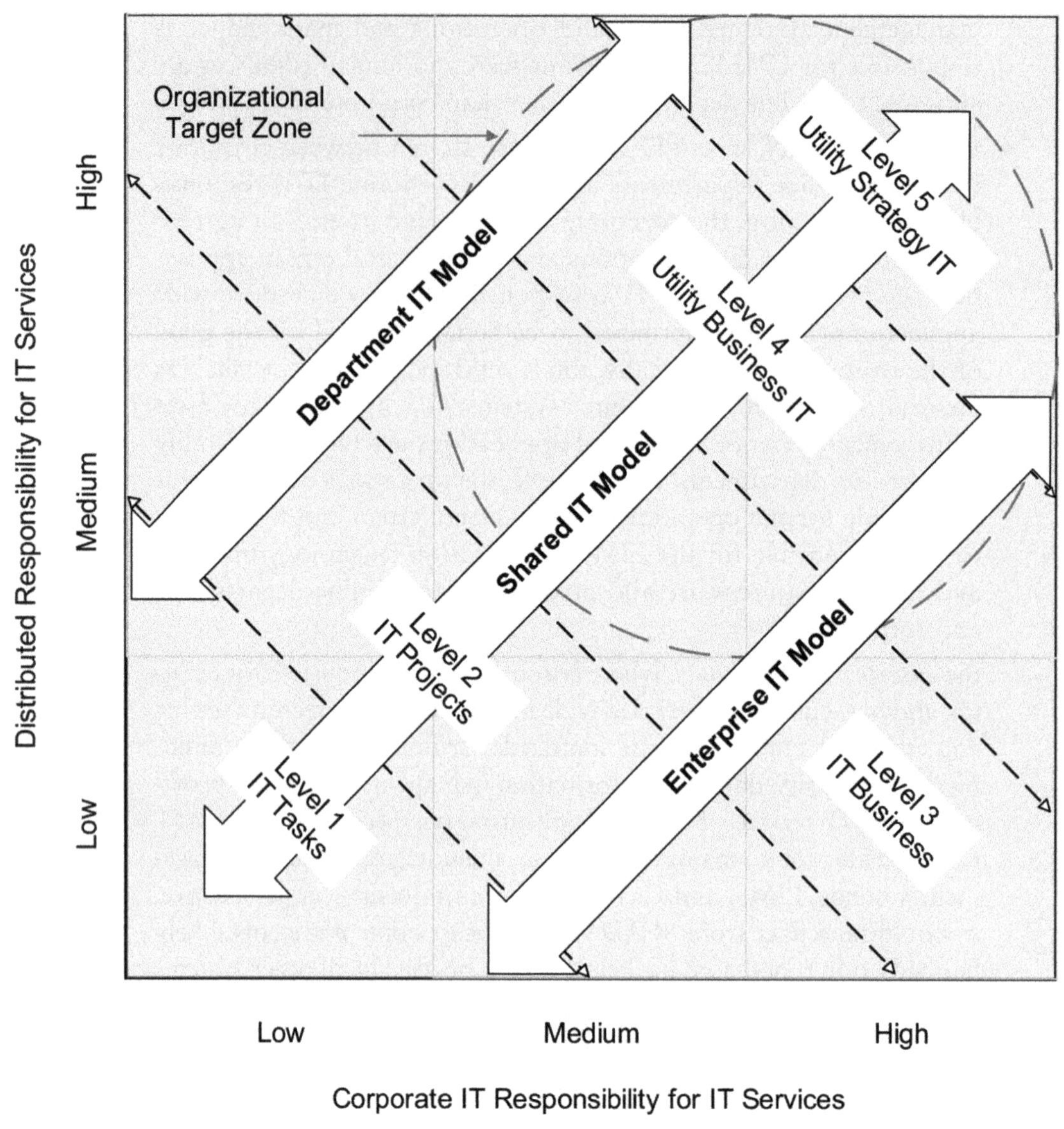

FIGURE 5.2 Map of Organization Options for IT in Utilities

IT organization. For each natural area of IT expertise, business solution delivery, and customer relationship management, the IT organization would establish teams responsible for centers of expertise, or IT hubs. The types of hubs that could be established include an architecture hub, a hardware environment hub, an e-utility hub, a customer/customer relationship hub, a business solutions hub, a project/program management hub, and a contract/contractor management hub.

The choice of how specific IT responsibilities are allocated across corporate IT, its hubs, and the rest of the utility would be based on a number of design factors. Examples might include

- internal capacity of IT at the department and corporate IT levels;
- effect of the selected organization on the ability to
    - provide for efficient and effective integrated process delivery,
    - accommodate new projects/programs,
    - leverage expertise (i.e., the same functions together),
    - provide a friendly customer interface, and
    - be responsive to customer requests and needs;
- manageable reporting structure (e.g., workload balance/reasonable number of reports);
- level of geographic customer distribution; and
- ability of the local IT service marketplace to provide resources.

One strategy followed by utility IT organizations to be closer to their customers is the allocation and colocation of IT associates to customer departments. In addition, specific customer-relationship managers can be assigned responsibility for building strong links to front-line departments.

4.5 Fill New Positions and Build Better Connections

There is a trend in utility IT organizations toward hiring staff who bridge business and IT perspectives. Attracting new employees with strong IT skills to utilities can be a challenge. Competitive salaries, systematic mentorship programs, explicit career path development, and use of the strong marketing allure of the environment and public service are all required to successfully compete for increasingly scarce IT resources. Meeting the challenge of paying outside traditional pay scales for municipal employees will have to progress. The future lies in the creation of municipal pay scales that recognize a technical stream and a management stream of job progression, both of which are calibrated to market conditions.

Figure 5.3 shows roles and interactions that would be characteristic of the mature utility IT organization of the future. It also shows a strategic IT governance council and its potential members, including the CIO.

Decisions on allocating responsibility for service delivery would ultimately drive which IT skills and talent reside inside the organization, where they would reside, and how they would interact with other IT resources. The IT organization of the future would respond to the utility's business needs

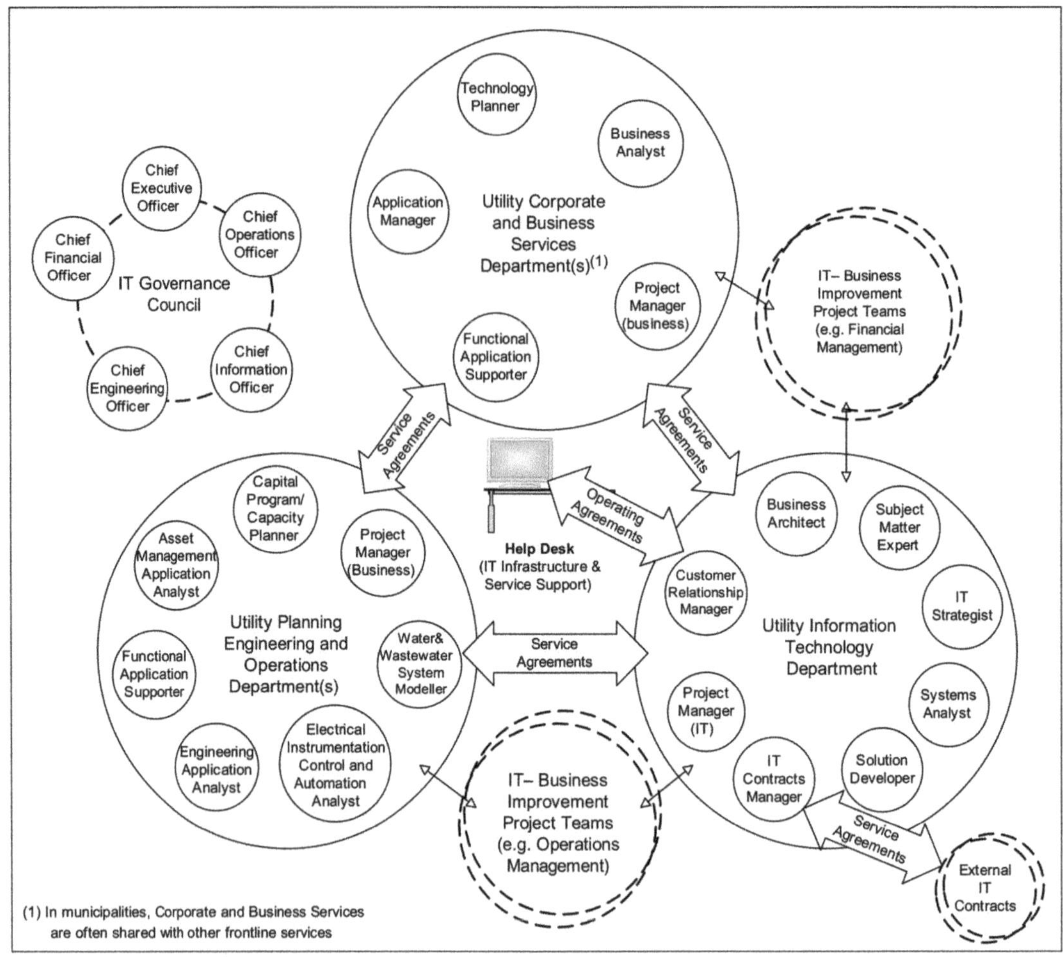

FIGURE 5.3 Roles and Relationships Map in the Mature Utility IT Organization of the Future

and make sure relationships with internal customers and external partners work well. It would then translate those business needs into information, application and technology architectures, integrated solutions, and IT service delivery processes while leading and facilitating the processes to connect with internal customers.

4.6 Get Strategic and Raise the Profile

It is emerging best practice for the utility's IT leaders to provide strategic advice as members of the utility's executive management team. In this capacity, they advise on how IT can contribute to addressing the utility's strategic business challenges, how new technologies could be applied to deliver increasingly valuable services to customers and stakeholders, and how the strategic application of technologies can best leverage applications and other IT resources for the benefit of the entire utility.

Utilities should establish an executive position such as vice president of IT or CIO who reports directly to the utility chief executive officer, general manager, or president. This demonstrates the utility's understanding of the importance of IT to all aspects of utility operations. In municipalities, they could report to the city manager or chief administrative officer. The IT leader in this position would have a full seat on the utility's executive team to allow ongoing strategic input into the business of the utility.

Secondly, IT leaders could establish an IT governance structure that would complement the corporate governance structure and include an IT forum under executive business sponsorship to govern the development and promulgation of all strategic utility-wide IT frameworks and standards. This forum, sometimes called an *IT governance council*, ensures that everyone adheres to the main principle of maximizing IT value for the entire utility enterprise.

4.7 Decide Who Does What

There are a number of models for defining decision-making responsibilities. Utilities exploring new governance and organizational structures should ensure that the model fits the culture of the utility and the organizational context within which the utility exists. Weill and Ross (2005) presented an IT decision-making matrix that includes six types of governance archetypes, five IT decision-making domains, and the opportunity to assign decision-making or input rights to each domain per group or individual. This matrix can help utilities clearly assign responsibility for making key decisions relating to IT. The archetypes include

- *business monarchy*, which is a centralized, business enterprise-centric model, in which responsibility for IT decision making and resource allocation is assigned to senior business executive(s) or an IT governance council for one or more key IT decision-making domains.
- *information technology monarchy*, which is a centralized, corporate IT-centric model, in which responsibility for IT decision making and resource allocation is assigned to the CIO and IT leaders for one or more key IT decision-making domains.
- *federal*, which is a shared model, in which responsibility for IT decision making and resource allocation is assigned to both front-line departments and the corporate IT department. Business-driven collaboration between the front-line departments and corporate IT drives the organization-wide optimization of IT and its role in the utility.

- *information technology duopoly,* which is a shared model, in which responsibility for IT decision making and resource allocation is assigned to department heads or commissioners and the CIO and IT leadership. Organization-wide optimization of IT is dependent on collaboration between the front-line departments and corporate IT.
- *feudal,* which is a decentralized, business-unit-focused model, in which responsibility for IT decision making and resource allocation is assigned to managers of business units or business processes.
- *anarchy,* which is the most decentralized model, in which responsibility for IT decision making and resource allocation is assigned to individuals or small groups.

The five decision-making domains are (1) IT principles and policies, which describe the high-level strategies that guide the way IT would provide most value to the utility; (2) IT architecture, which describes the overall framework for technologies, standards, and specifications that provides context for all IT systems; (3) IT infrastructure, which describes the specific hardware and communications infrastructure required to provide access and information sharing across the utility; (4) business application needs, which describe the business needs and related application software capabilities required to run the utility; and (5) IT investments, which describe how much money will be invested in which part of the service or organization. Each of these decision-making domains can be assigned to the enterprise, department, business unit, or group, and the individual levels.

Weill and Ross (2005) proposed that stakeholder groups or individuals be allocated either input or decision rights in each domain based on the aforementioned archetypes.

4.8 Manage the Change

Managing change well is the final step for utility IT organizations to undertake to ensure a successful transformation from the current as-is state to the target to-be maturity level, organizational structure, and governance framework. Although there are many ways to manage change, the process is often cut short for expediency reasons or lack of available change-management skills. A recent survey of the UK public sector highlights the shortage of change-management expertise as being the number one skills-related hurdle to improving municipal performance.

Successful changes are those that make certain all important change elements are cared for in a timely manner. A compelling vision would have been in place, a shared sense of urgency would have fueled the journey, visible leadership would have guided the process, a clear plan would have been followed, and the appropriate resources would have been applied along with incentives to deliver in a reasonable time frame. The change equation in Figure 5.4 shows elements that govern a successful transformation. It is important to recognize that if any of the elements in the numerator in the equation is zero, there is no change possible.

There are a number of examples of methodologies that can be used to guide the change process. Some find their roots at the personal-change level, like the best-selling *The 7 Habits of Highly Effective People* (Covey, 1989) and *ADKAR: A Model for Change in Business, Government and Our Community* (Hiatt, 2006). The former is based on the "maturity continuum" of personal and interpersonal effectiveness and on the development of habits created by combining knowledge, skills, and desire; the latter combines personal and professional awareness, desire, knowledge, ability, and reinforcement.

Other change models find their genesis in IT project methodologies such as Prince2, which uses the following five-step process: (1) set a common purpose, (2) mobilize resources, (3) plan and design the solution, (4) implement the solution, and (5) secure the result. Prince2 is a process-based approach for project management, providing an easily tailored and scalable project management methodology for the management of all types of projects. It does not spend much time on the underlying motivations of people but is focused on delivering the objectives of the project. The method is the

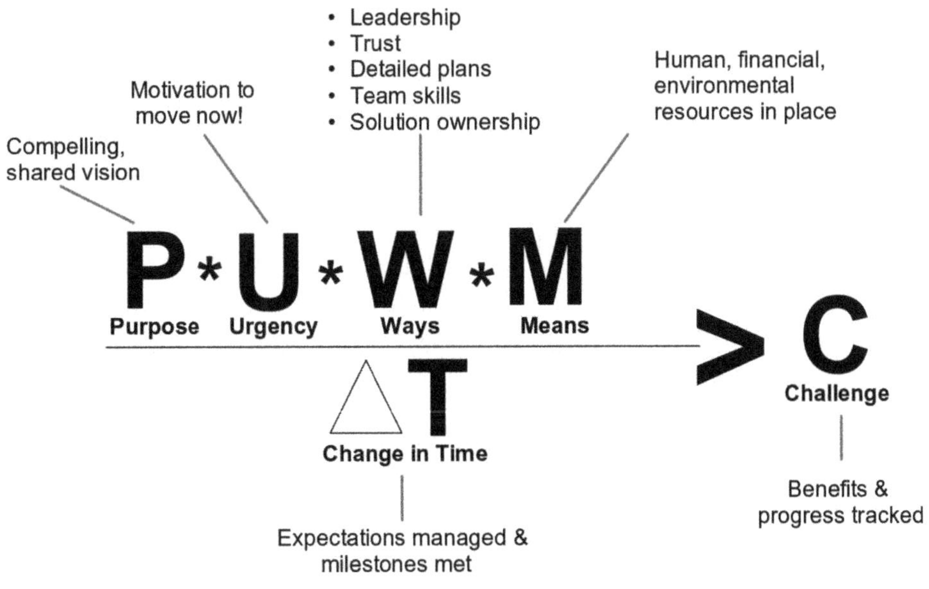

FIGURE 5.4 Change Equation

de facto standard for project management in the United Kingdom and is practiced worldwide.

Yet other change models are aimed at transforming organizations based on vision, people, culture, and results. For example, *Leading Change* (Kotter, 1996) and the follow-up *The Heart of Change* (Kotter & Cohen, 2002) outline reasons why change processes fail. The books describe eight phases of change as follows: (1) establish a sense of urgency, (2) create a guiding coalition and guiding teams, (3) get the vision right, (4) communicate the vision for buy-in, (5) enable action and empower people to clear obstacles, (6) create short-term wins, (7) don't let up and keep moving, and (8) make it stick.

All change models can help, and the best one should be selected based on the type of challenge the utility IT organization faces. Regardless of the chosen model, success requires a level of stakeholder engagement and communications commensurate with the nature of the challenge.

This engagement and communication should occur in a strategic manner, captured in a framework that poses the following questions:

- Which are the highest priority stakeholders and how much effect do they have on the success of the change?
- What are the needs and interests of the priority stakeholders related to this?
- What communication and engagement strategies should we follow for each stakeholder group?
- Want messages are appropriate for the group?
- What specific methods should we use to communicate?
- What specific events would be best and how frequently should they occur?
- Who should be responsible for communicating?

The final element to manage well in the change process is development of the culture. Often, there are opportunities for utility IT organizations to adjust their culture to better align with the culture of the overall utility organization. This can be done by basing behavior and decisions on a set of clear values and principles and becoming more customer oriented and responsive. There are three powerful drivers that result in true culture change and transformation in utilities. These are

- the "burning platform" of fear, risk, or disaster, such as in Milwaukee, Wisconsin, where a cryptosporidium outbreak in 1993 brought in the need for significantly improved quality monitoring and treatment management while putting the brakes on a trend to transform to "city-as-a-business" governance constructs that had been brought in by Mayor John Norquist in 1988;

- the "wolf-at-the-door" prospect of competition, such as the entry of private-sector water and wastewater operators into the North American marketplace in the late 1990s and early 2000s; and
- the "nirvana" of being the best, such as utilities in Colorado Springs, Colorado (e.g., Colorado Springs Utilities), and Edmonton, Alberta, Canada (e.g., EPCOR), which both quickly transformed from responding to competitive pressures to focusing on great results and having award-winning governance frameworks in place.

The first two drivers are more likely to provide the momentum for change as the lack of true carrots and sticks in the public sector make it difficult for leaders to create the necessary sense of urgency. The third driver is the aspirational motivator that delivers the most sustainable culture change but is the most difficult to achieve.

The IT response to these cultural change drivers should be swift and accurate. If the utility's challenge is related to a number of environmental disasters caused by poor management of treatment systems and sampling information, IT should be there to help with technology solutions. This requires an IT culture that includes a mix of strategy, customer-service orientation, responsiveness, and utility business awareness. In addition, if the utility is under stress from outside influences, the IT organization needs to be sensitive to the effect of that situation on its customers and staff.

The utility IT organization itself has a similar set of drivers. Offshoring of services; availability of services through local consultants, contractors, and vendors; and the increase in the number of application services that can be provided outside the corporate firewalls all provide competitive challenges. These drivers can provide the impetus for IT to change its culture to be more flexible, business oriented, and focused on the well-being of the utility. In addition, IT governance has an important role to play in setting the right philosophy and direction and managing integrated IT service delivery performance. The governance vision for leading utility IT organizations in the future should include recognition by industry associations.

5.0 CULTURE AND TRUST

The success of any organizational and governance system is dependent upon the organization's culture and level of trust. The guidance and recommendations in this manual of practice support the development of processes, procedures, roles, responsibilities, and documentation that will reflect the organization's culture, and that provide a framework for having transparency in decision making and setting clear expectations.

When there are difficulties associated with IT, having a culture that supports teamwork, learning, and empathy will enable quicker resolution than one that is divided. Of course, this is true of all situations, not just IT-related activities. However, historically IT has been a common place where culture and trust issues emerge. This may be because when enterprise IT systems were first introduced in the industry they were presented as efficiency measures that could reduce the number of staff. The result was skepticism and in some cases alienation of the staff who were critical in ensuring that the tools were used properly. Another contributing factor is that the language and terminology of IT is different than that used by utility staff, resulting in misunderstanding and creating divisions.

Utilities that feel that their IT operations are not well aligned with the organization's priorities could consider evaluating the culture, including trust, communications, and teamwork. Change management programs that are put in place when there is a big initiative include some components of this but are typically focused on managing a specific source of change rather than addressing the underlying culture. Proactively addressing culture before initiating a change initiative can not only lead to better outcomes for large projects, but can also increase people's satisfaction with their jobs, leading to better outcomes for the utility and customers.

6.0 REFERENCES

Covey, S. R. (1989). *The 7 habits of highly effective people.* Simon & Schuster.

Hiatt, J. M. (2006). *ADKAR: A model for change in business, government and our community.* Prosci Learning Center Publications.

Kotter, J. P. (1996). *Leading change.* Harvard Business School Press.

Kotter, J. P., & Cohen, D. (2002). *The heart of change* (1st ed.). Harvard Business School Press.

Salkowitz, R. (2008). *Generation blend: Managing across the technology age gap.* Wiley & Sons.

Weill, P., & Ross, J. W. (2005). A matrixed approach to designing IT governance. *MIT Sloan Management Review*, 46(2), 26–34.

7.0 SUGGESTED READINGS

Baldoni, J. (2003). *Great communication secrets of great leaders.* McGraw-Hill.

Rau, K. G. (2004) Effective governance of IT: Design, objectives, roles and relationships. *Information Systems Management*, *21*(4), 35–42.

Seppala, E., & Cameron, K. (2015, December 1). Proof that positive work cultures are more productive. *Harvard Business Review.*

Shapiro, R. J. (2008). *Futurecast: How superpowers, populations, and globalization will change the way you live and work.* St. Martin's Press.

6

Information Technology Capital Project Management

1.0 SUMMARY OF KEY THINGS TO KNOW

Project management best practices are relevant to information technology (IT) projects as well. The intent of this chapter is not to imply that different practices be applied. Rather, it is to highlight those aspects of IT projects that may require more attention than other types of projects such as construction projects. Changes to IT invokes change on all processes, policies, and staff associated with the system. Alhough utilities now have many years of experience with IT, technology has continued to evolve rapidly and will continue to do so, resulting in a different set of expectations and concerns in the project management process.

- Project Management Institute's (2021) *Project Management Body of Knowledge (PMBOK) Guide* is the standard of practice for IT project management.
- IT projects have different considerations than capital projects involving physical infrastructure.
- IT initiatives can be structured as a program made up of shorter projects to mitigate risk and deliver outcomes sooner.
- IT projects should be planned and managed in alignment with related utility priorities that they are supporting.
- Users are critical stakeholders and should be part of the project team.

2.0 INTRODUCTION TO INFORMATION TECHNOLOGY PROJECT MANAGEMENT

Implementations of complex initiatives in all fields of human endeavor share a common set of pathways to failure that can foil the best laid plans. These pathways to failure arise in several dimensions and require active intervention to keep an initiative on track to successful implementation. These are discussed here and also listed at the end of this chapter as a "project management checklist" for reference.

Complex initiatives range from activities such as constructing a new facility, undertaking an asset management program, or merging with another organization. Complex initiatives are various forms of projects or programs. Programs are simply logical groups of projects.

Managing projects and programs has become a specialized science and structured around the approximately 12 dimensions that can affect success. Some of these dimensions are obvious (e.g., time, scope, and money). Other dimensions might be less obvious and require some thought. The

Project Management Institute's Project Management Professional (PMP) certification has become the standard across industries to educate and train professionals tasked with overseeing projects and programs. The certification covers the many dimensions of project management by explaining the levels of control to guide projects from concept to implementation. There is a common perception that IT projects are always late and over budget. In some instances, the problem is poor management of expectations, and the perception is a result of poor communication management. In others, the problem is poor management in several dimensions and the perception is the reality. Good project management keeps the project on a course to successful implementation and allows the project manager to identify and avoid the pathways to failure in multiple dimensions.

2.1 Project Management Challenge

The importance of the project management function is often underestimated, and for some time the role was assigned without adequate training on the complexities of the responsibility. The first challenge is to have the necessary support and expectations for the project manager, their tools, and their team. This includes following PMI standard processes.

A second project management challenge is having the technical implementation team members think that project management is not really needed. Information technology specialists, scientists, and senior management can individually see and discharge their technical scope functions but might be unable to see that there is need for a global organizational framework around a project or that they need to defer to such a framework. Therefore, they might also conclude that rigorous project management is not really necessary for success.

A third project management challenge, and a fatal flaw for a project, is in not budgeting for the project management function at an adequate level to ensure that all the dimensions of project management are applied throughout the project's life cycle.

The project management challenge can thus be summarized in terms of sponsorship for project management, team building around an explicit project management structure, and adequately resourcing the project management function.

2.2 Information Technology Design Challenges

Information technology is a rapidly evolving area as humans discover new ways of acquiring, storing, retrieving, managing, and manipulating data and information. Each of the subtechnologies of IT is a science in its own right, and practitioners are continually specializing into narrower aliquots of their fields as they drill deeper for learning and develop applications of the new knowledge. Therefore, it is understandable that IT practitioners are generally more focused on the technology itself than on the surrounding issues of a project. Information technology has more evolution ahead as it resolves into the science and the application of IT.

With rapid proliferation of technology approaches, development of special jargon, early-to-market applications, and so forth, design of an IT project poses unique challenges. Someone has to understand the IT concepts and also understand project management concepts and bridge the communication gaps between sponsor, technical team, and other stakeholders, notably users. It seems that the universe of IT knowledge is expanding far faster than bridge builders are entering the field to support this expanding universe. This can lead to the following two undesirable outcomes: proceed without proper project management or defer the project.

A particularly important design challenge for an IT project is in bridging the communication gap between users and vendors and other IT professionals. Users know the underlying business processes and are best positioned to understand current and future business needs. However, users are generally not knowledgeable about how business processes are supported by IT itself. This is analogous to the pilot of an airplane understanding the aerodynamics of flight and navigation, but not having detailed knowledge of exactly what is happening when the throttle or power lever is moved and the engine is required to increase or decrease power output. Users are pilots; IT designers are engineers. Users/pilots know what performance envelope is required; IT designers/engineers develop the underlying machine to achieve the required performance envelope. The challenge is to find a common language for them so that expectations can be set, met, and demonstrated before the system goes live. Users have increasingly been exposed to IT terminology on the job and in their personal lives, yet it is critical to find a common language and reduce jargon when communicating across a diverse team.

2.3 Obsolescence Challenge

Rapidly evolving systems generate obsolescence. Practically all IT systems implemented by a utility will require ongoing upgrade through their economic lives to remain useful within an interconnected architecture. It is not sufficient that a technology subsystem continues to function well in its core duty alone; it must increasingly communicate with newer technology subsystems in ways that could not have been anticipated at an earlier state of knowledge. There are two major ways for a program to become obsolete before the end of its anticipated economic life cycle: (a) changes in the required functionality or (b) the emerging need to integrate with other computer systems or components.

For example, a supervisory control and data acquisition (SCADA) system might continue to operate a pump satisfactorily for decades after initial implementation, but the system might not be able to pass on information about how many running hours the same pump is accumulating unless the interconnectivity is compatible. If the SCADA system was originally designed only to store run hours in its proprietary time-series database for display on the SCADA screen, then that would be the limit of its capabilities. The program manager should consider that water utility business processes might have since evolved to integrate such data into other systems, such as asset management systems, enterprise resource planning systems, and so forth, and that the new connectivity requirement is to pass on information for a thousand operating assets and to do so wirelessly, at a high rate over long distances. As such, the SCADA system might now be obsolete.

All IT projects should be approached with the concept of ongoing requirements for upgrades. This has implications for management of the system development life cycle (SDLC). Project managers should be aware that time is really of the essence with these projects and that time management has two imperatives from an SDLC viewpoint. First, the program manager should capsule user requirements and expeditiously move the project to completion with minimal changes and enhancements along the way; second, the program manager should prepare the way for commencing a subsequent SDLC immediately upon completion of the current one to deal with desired changes and enhancements that came to light during the current SDLC.

Sponsors, stakeholders, and users are generally unaware that time management is so critical in IT projects compared to conventional engineering projects. This is a major mode of failure for IT projects with their high obsolescence challenge.

3.0 OVERVIEW OF PROJECT MANAGEMENT METHODOLOGIES FOR INFORMATION TECHNOLOGY

Good project management methodologies have common elements: structure, procedural clarity, input and reporting points for sponsor control, documentation of process to enable learning, flexibility to manage the unexpected, and so forth. In this chapter, the authors will refer to certain "actors" relevant to project management. As such, these actors are defined in general terms, as follows (it is important to note that specific projects might use variations to these definitions to suit the vocabulary and culture of the project context):

- *project manager*—the person responsible for executing delivery of the project. This person is the ultimate authority governing activities of the project team. The project manager reports to the sponsor.
- *project team*—the persons who, individually and collectively, have responsibilities to deliver each task in the project work breakdown structure.
- *project sponsor* (sometimes called the "*project champion*")—the individual who takes organizational and political responsibility for achieving project goals and enables the necessary budgetary resources to sustain the project. The sponsor is often the key to managing external stakeholder relationships that are beyond the reach of the project manager.
- *client*—the person or organization receiving the benefit of the completed project and often the source of funding for the project.
- *users*—the individuals who will accept the project on behalf of the client and operate the system after delivery.
- *stakeholder*—any person or organization who has a valid interest in the project.

3.1 Program Versus Project Management

In this chapter, the terms "program" and "project" are used interchangeably. Program will generally be used as a collective term for a cluster of projects, in the same way that flock is a collective term for a cluster of sheep. Although other meanings exist for the term, such other uses will be clarified in the context of this chapter.

It is often convenient and desirable to cluster individual projects into a single initiative (i.e., program) within a business environment for reasons varying from commonality of objectives, obtaining economies of scale, serving multiple stakeholder requirements, and so forth. Hence, it is quite common to see IT capital programs made up of several projects.

In the context of the practice of project management, the term "*project*" is also used as a collective term by the Project Management Institute (PMI) (Newtown, Pennsylvania), a widely recognized professional organization that has created *Project Management Body of Knowledge (PMBOK) Guide* (2021), which represents a body of knowledge that codifies the discipline of project management. This discipline applies to individual projects and, equally, to programs.

3.2 System Development Life-Cycle Model

Project life cycles typically follow the classic "S" curve found throughout nature, describing such processes as transmission of infections in a population, dissemination of information into a public, cumulative work effort over a project, and so forth. These phenomena typically have a starting point and typically taper to an end point.

In the SDLC "S" curve in Figure 6.1, the approximate end of each phase is shown by the arrows marking definition, design, development or configuration, and release. These phases are described in more detail later in this chapter.

The SDLC model for IT projects is best portrayed as a series of life-cycle curves, as illustrated in Figure 6.2, to emphasize that *time* is of the essence. Those desirable changes and enhancements identified during the current development are typically best deferred to a subsequent cycle (or "release" or "version") to maintain focus on the initial user requirements that drove the project at kickoff. This race to completion is primarily driven by the risk of obsolescence discussed previously, although there are collateral benefits to maintaining a pragmatically narrow focus relating to managing scope creep, resource usage, efficiency, and so forth. The SDLC for IT projects has clear phases as described in the following sections.

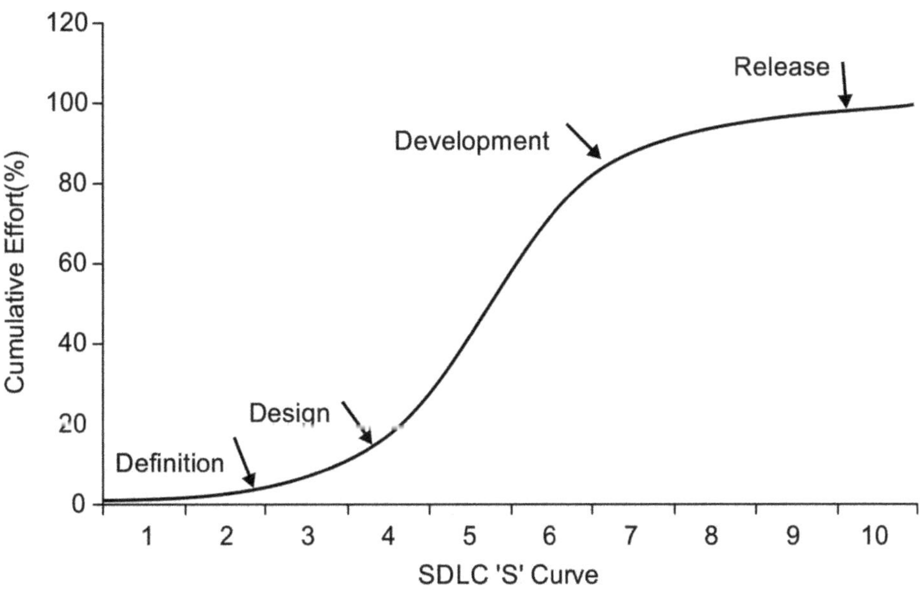

FIGURE 6.1 System Development Life-Cycle "S" Curve

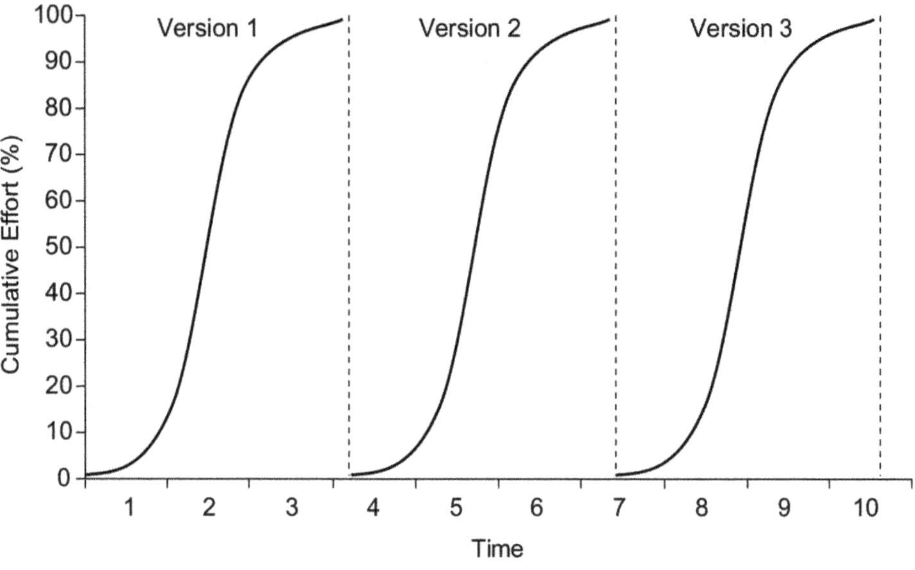

FIGURE 6.2 System Development Life-Cycle "S" Curve (With Upgrades and Enhancements Subsequent to Version 1)

3.2.1 Definition

During the definition phase, the project is managed through certain key activities and milestones. These include identification of need; preliminary study to frame the scope, schedule, and budget; identification of resources; empowerment of a sponsor; and authorization to undertake the project. This phase typically ends at about the time the project sponsor issues a kickoff directive.

3.2.2 Design

During the design phase, the project assumes more structure and involves increasingly larger numbers of people. Often, operational staff in a water utility will turn over the project to an internal or external project manager to build and maintain the necessary project management infrastructure. Activities and milestones of this phase include the following:

- Set up a project management office (PMO).
- Define the core team of staff supporting the implementation.
- Prepare user requirements.
- Prepare a predesign report.
- Report to stakeholders and sponsor to receive a go/no-go decision before proceeding to the next phase.

Each of these activities has associated detailed tasks, such as setting up the 12 dimensions of management inside the PMO, convening user workshops to extract and document user requirements in plain language, procuring consulting resources to supplement in-house resources where necessary, and so forth. This phase typically ends at about the time that the predesign report has been signed off by users and accepted by the PMO.

This phase most commonly involves preparing for the selection of a commercial software product from an established vendor. The predesign report, including the user requirements, will become the basis for a request for proposal from software vendors and will become the basis for the evaluation of responses. Additional information about software procurement is presented in Chapter 7.

3.2.3 Configuration and Development

During the development phase, the project builds momentum exponentially with time. The development phase is the most intense, costly, and complex phase of an IT project. Development typically commences following the selection of a software product and implementer, with expansion of the predesign report to a detailed design brief, during which several important deliverables are generated. An early deliverable is conversion of the user requirements into technical language, thereafter known as *user specifications*.

Implementers will use the specifications to configure the software to meet the utility's needs. If there is a critical business need that is not part of the selected software, the specifications will be used to define the required customization. It is not advisable to include customizations unless absolutely necessary because they introduce risk into the implementation plan, and they can be expensive and time consuming to maintain.

The implementer will work with the project manager to define the work breakdown structure to break the project into components that can be assigned to individual task owners. Another deliverable is documentation of the testing protocols and plans for unit testing of the small components as they are developed, followed by integration testing of the components as they are fitted together. An important phase-end deliverable is acceptance testing by the users before allowing the new system to go live. Testing is a good opportunity to begin to get key staff comfortable with the new system(s). The test scripts should be based upon the utility's business processes and provide an early training opportunity for staff to understand what will be different in their work activities with the new tools.

At each major milestone of the development stage, appropriate reporting is provided by the PMO to the sponsor and other stakeholders to ensure ongoing support is maintained through the inevitable challenges of dealing with unknowns as they arise, an unfortunate characteristic of all projects. Throughout the project, the PMO's mandate is to continually manage the potential pathways to failure discussed previously; these are in the dimensions of time, scope, money, quality, resources, communication, change, risk, data, procurement, standards, and integration. Integration here means the overlap of the other dimensions of project management, including trade-offs and balance among combinations such as time and money, money and risk, change and risk, and so forth.

3.2.4 Release

During the release stage, system design has been tested and accepted by the user group and is ready to be used as a production system. Project management functions at this stage primarily focus on time, scope, and money. The technical resources, meanwhile, focus more intently on the schedule and cutover to an operational system, with rapid response to bugs that might not have been caught in testing before going live. The technical team leads the project through release, and the PMO monitors and documents progress while maintaining the project support infrastructure, such as procurement management of any contracted resources. By this stage, assuming good project management in prior stages, the major work is largely done, budgets are mostly expended, and wrapping up the project consumes the PMO's attention.

3.3 Why Structure Is Necessary

The complexity of many IT projects can be enormous and often cannot be fully understood by stakeholders without a complete model of the project being prepared and documented through such tools as Gantt and program evaluation and review technique charts, work breakdown structures in nested hierarchies, resource assignments, and so forth. Just managing data in a major project might require development of a document management system, which, in itself, is a project within a project.

Coupled with this technical complexity is the apparently even more puzzling complexity of human interactions with personality and emotional overtones that can make a team underperform. Some of these interactions and relationships are portrayed in Figure 6.3, which shows a model of project management centered on a PMO and bidirectionally focused on external stakeholders and the project team, respectively.

Finally, it is important to consider the business context within which the project resides. The business itself might be under stress that is either

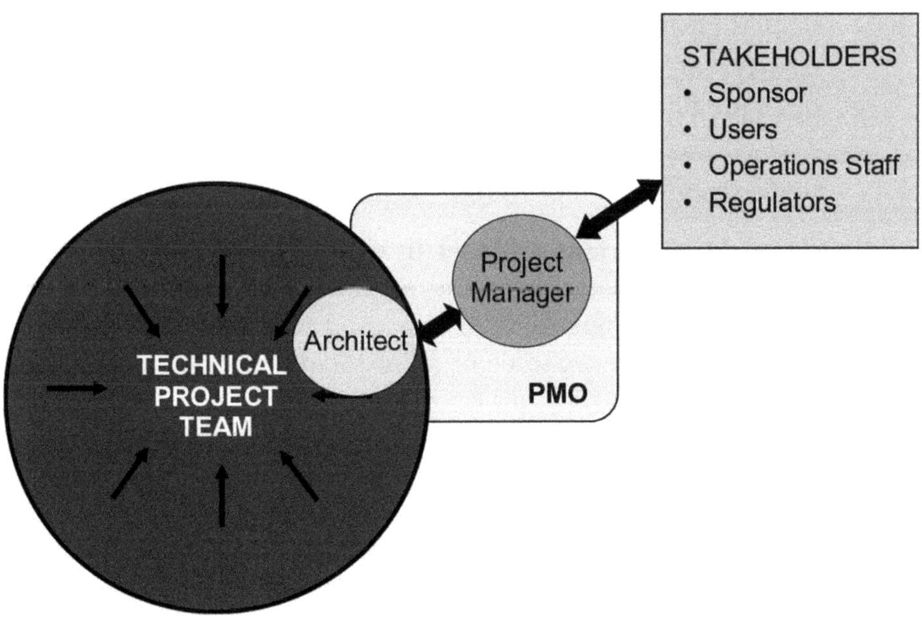

FIGURE 6.3 Project Management Model

related or unrelated to the project; moreover, constant competition for resources requires a sponsor to guard the project throughout its life cycle. The project manager needs the sponsor's continuous support to ensure that circumstances beyond the project manager's authority are placed before the sponsor for attention and action.

A properly orchestrated PMO allows each of the elements of the project to play out and guides the project to ensure that it stays on track despite a constant tendency, common to all projects, toward instability and failure. Without the structure of such a PMO, a complex project has little chance of meeting all its objectives.

For practical examples of project plan structures for IT projects, the reader is encouraged to seek out readily available resources on the internet using any of the popular search engines.

3.4 Other Methodologies

There are other ways to manage projects that might be appropriate in certain circumstances. However, the PMI-based methodology described previously, and variations around it, constitute mainstream ideas for project management of IT projects for water utilities. Specialized methodologies are beyond the scope of this chapter.

4.0 GUIDELINES FOR MANAGING WATER AND WASTEWATER UTILITY INFORMATION TECHNOLOGY PROJECTS

4.1 Understanding Multirational Organizations

Multirational organizations are ones in which several stakeholders or groups of stakeholders have divergent opinions on the best way to solve a problem. Generally, these opinions are valid when seen from an individual perspective; however, some or all of them might be inadequate to truly solve the problem. The project manager acting in a multirational organization is challenged to develop solutions within a context of strong-willed stakeholders whose buy-in and participation in a final solution is vital for success.

There are numerous examples of multirationality in the political arena (e.g., the European Union, the senates and parliaments of democracies, and so forth). Multirationality also abounds in corporate entities with consensual cultures; water utilities can be among such entities.

As an exercise in understanding multirationality, consider the project management challenge in supporting the technical team as a workshop convened to define user requirements. Assume that the user community is made up of a water operations group and a corporate financial management group (add more groups to fuel the fire in this analysis). Further assume that the culture of the water utility has recently undergone reorganization and that members of the two groups hardly knew each other before the project was launched.

Now attempt to develop a common set of user requirements for a new system, such as an asset management system, a SCADA system, or another system. Each subgroup of this newly convened user group would have come to the table with a set of notions influenced by the historical behaviors of the subgroup, its cultural mores, its asset management structure, its financial and accounting structure, and so forth. Each subgroup's position on user requirements would be completely valid from its own perspective but might not be valid in support of the terms of reference of the new project.

To make progress within a multirational organization toward developing a common output such as a common set of user requirements, it is necessary to build a team that looks beyond the initial positions and toward the common goal. To build such a team, much preliminary work would need to be undertaken to validate the initial positions and to provide recognition of the gaps between the initial positions and the desired new project goals. Such is the nature of multirationality.

4.1.1 Managing Sponsors

Managing the sponsor is directed toward one purpose: communicating clearly to allow informed decision making by the sponsor where these are beyond the authority level of the project manager. A sponsor must always exist for a project to succeed. Some sponsors prefer to remain anonymous for political or other reasons and insert a proxy sponsor to front a project. Although this can be effective, it makes the project manager's work more difficult because clear communication is less likely to occur.

It is suggested that program managers define *sponsor* as that individual or group of individuals who have high-level risk exposure from the failure of the project and who have the authority to initiate the project and to provide resources for its execution. The first element is important; if a sponsor cannot be found who cares about project success and who suffers potential risks from its failure, the project is likely to fail because all projects inevitably hit rocky spots where a sponsor will be required to clear the way for the project manager.

For example, "Project A" has been initiated in a department of the utility and is midway through implementation when a challenge arises from the head of another department of the utility who would like to sponsor "Project B" to achieve the same objectives, but with a different political slant. If Project B gains momentum, the manager of Project A would have little ability to resolve which project should prevail. The sponsor of Project A would need to coordinate with the sponsor of Project B to reach consensus on how best to proceed. The project manager's responsibility for Project A would be to bring to the sponsor's attention any issues relating to the project that are beyond the authority level of project management and to seek continuing support and sponsorship for the project.

4.1.2 Managing Users

Users are the closest stakeholders whom the project team has relationships with. They are critical from the project manager's point of view because they are responsible for outlining functional requirements and for accepting the final deliverables produced by the project team. Users set expectations for the performance of the product at project initiation and are important as acceptance "gatekeepers" for the product nearing project end.

The project manager's relationship with the user group requires careful attention. Because users are not typically "technical" in the IT design sense, they may not be able to specify in technical terms how they want the product to perform, although they do know their requirements. Thus, the project manager must ensure that there is an accurate interpretative stage when user requirements are translated from plain language into technical language.

More importantly, the project manager must ensure that the user group and the technical team understand the function of the user group as gatekeepers for acceptance. Early achievement of this understanding will set the stage for avoiding several pitfalls relating to user acceptance. Examples of pitfalls include failure of users to set comprehensive requirements up front, failure to validate user requirements and achieve sign off for the project scope (no more and no less), failure to interpret user requirements accurately, failure to design according to the real requirements, failure to enforce acceptance of properly designed solutions if users change their minds as to scope, and so forth.

The project management office needs to explain to users and the technical team the potential failure modes and ensure that all parties understand the important roles they play in progressing toward a successful project. In addition, the project management team should include representatives from the user community in key project meetings and decisions. This has the benefit of transparency and ownership of outcomes with the user community and provides users with additional understanding that can help with user training.

4.1.3 Managing Technical Experts

In terms of managing technical experts, the important project management skills lie in the area of resource management. Technical issues are the domain of the technical project lead or architect, and the PMO should avoid these issues. However, technical personnel in general have less tolerance for team relationship issues and can, therefore, trend toward low performance when there is real or potential discord on selecting appropriate technical solutions. A team-development model ("forming, storming, 'norming,' performing") initiated by Tuckman (1965) is a useful construct for these situations and is briefly described here.

Experience has shown that, in many situations, formal team-building workshops pay significant dividends in trending toward high-performance teams. Left to themselves, teams tend to go through four common phases: forming, storming, norming, and performing. These phases represent successive stages in relationship building. "Forming" is Stage 1 and occurs when two or more people assemble for the purpose of undertaking a common goal. "Storming" is Stage 2 and occurs as people in the group discover irritations among themselves (i.e., bickering) that distract them from working on the tasks necessary to achieve the common goal. "Norming" is Stage 3 and occurs as participants in the group start to understand and negotiate their dissonance into normalized acceptance of irritations or agreed changes to behaviors. "Performing" is Stage 4 and occurs after the team is effectively normalized and participants then return to concentrating on executing the tasks required for achieving the common goal. The key here is that although there is no shortcut or way of bypassing the intermediate stages, with facilitation the team reaches the performing stage sooner than without facilitation.

Without the investment in team-building workshops, teams are left to transition progressively through these phases on their own. However, there is a possibility that a team could become stuck in the storming phase and fail to normalize and graduate into the performing phase.

An experienced project manager must be vigilant in monitoring so that the technical team achieves high performance and must also intervene when it is appropriate to do so when teams are not working well.

4.1.4 Managing Vendors

Managing vendors lies in the project management skill set of procurement management. Vendors in IT projects include individual contractors, personnel agencies, suppliers of shrink-wrapped software, hardware suppliers, system integration developers, and so forth.

Procurement management includes ensuring that appropriate contracting and technical documentation is used, quoted prices are managed, invoices are promptly paid, and so forth. A purchasing department within the utility often provides valuable support to the PMO in this area, although, occasionally, the project requirements for a major program might demand quicker delivery than a major purchasing department can deliver. In such cases, the PMO could request delegated authority through the sponsor for alternative arrangements that would satisfy policy and audit requirements while speeding up the procurement function.

5.0 STRUCTURE OF INFORMATION TECHNOLOGY PROJECTS

5.1 Define, Design, Configure/Develop, and Release Phases

As discussed previously in this chapter, define, design, configure/develop, and release phases are logical segments for a work breakdown structure; they also represent logical go/no-go decision points for sponsor reaffirmation of need for a project. At each stage, more information becomes available in the key areas of time, scope, and money that are interesting to the sponsor. As information relevant to the time, scope, and money "basket" evolves, the sponsor has the opportunity to assess the project strategically and can guide it for best fit to strategic needs or cancel the project if future outlook warrants.

These four phases have individual project management requirements. The define phase is concerned with broad-scope views, the design phase is concerned with reducing uncertainty, the develop phase is concerned with details, and the release phase is concerned with final delivery and effect on the utility's operations.

5.2 System Development Life Cycle

The key concept conveyed by the SDLC is that IT projects are amenable to cyclical iteration; thus, they do not have to be totally comprehensive at the outset. Users can then tolerate misgivings they may have that the user requirements are not sufficiently comprehensive. Otherwise, user behavior will trend toward developing the user's requirements in too fine detail and without appropriate timeliness; in other words, it will become an interminable exercise.

The cycle is intended to generate a Pareto level of functionality—capture, say, 80% of the required functionality before freezing the user requirements because the other 20% might not become apparent until well after the details have been developed. The cycle allows for efficient capture of the remaining 20% during the project and putting these into a basket of desirable enhancements to be implemented in the next cycle of the SDLC.

Some users have difficulty with this concept and argue that a project that is moved into production with only $x\%$ (where $x < 100$) of the functionality identified is unacceptable. If the corporate culture prevailing in the utility cannot support an SDLC approach, then the project manager must choose another methodology and ensure that, if longer project duration results from an alternate methodology, the stakeholders accept such an outcome. As mentioned earlier in this chapter, time is of the essence with IT projects because of the potential for obsolescence, which represents a real risk for project failure.

5.3 User Participation

User participation in an IT project historically occurred near the beginning and end of the project. However, standard practice is now to include users across all key project activities and to have them as part of the project team. Users are vital for developing requirements, understanding compromises required among conflicting requirements, agreeing to a testing and acceptance protocol, and participating in acceptance testing (i.e., asking, "Does the product satisfy the signed-off requirements?").

Once user requirements have been signed off on, users may not have a hands-on role in the project until acceptance testing is required. However, opportunities can be built into the schedule to have users' support with data migration, interface configuration' and training material development. This prevents the problems that may occur over long projects where staff change, and users lose connection to the project. Hence, it is important to charter the user group and have a governance structure for the long term.

5.4 Testing and Acceptance

Testing is formal activity that helps satisfy the PMO's requirements under quality management. There are various types of testing that occur in the life cycle of an IT project; unit testing occurs as programmers develop code; internal acceptance testing is done by the technical lead or architect; integration testing occurs as modules are fitted together; and, finally, acceptance testing is done by the user group before going live.

Unit testing is performed as part of system configuration and development and is undertaken by the implementer or programmer, typically the software vendor's representative. The objective of unit testing is to show that the system works and satisfies the specifications.

Internal acceptance testing is done by designated members of the implementation team other than the person who configured or wrote the code. Unlike unit testing, which is geared toward showing that the code does work, internal acceptance testing is geared toward finding conditions under which the code does *not* work; hence, the requirement for independence for the internal acceptance tester from the code designer. If the system fails internal acceptance testing, the module configuration or development is reiterated until it passes both unit testing and internal acceptance testing. This iteration is common and should not be viewed with dismay by the PMO; rather, some allowance should be built into the project to accommodate a few reasonable iterations.

Integration testing is conducted within a special testing environment, that is, a computer system separate from the development environment and closely mimicking the final production environment. Modules are migrated from the development environment for testing in the testing environment. As modules develop, they are tested for compatibility with other modules in accordance with a testing protocol designed by the architect. If a module fails an integration test, it is returned to the development environment for reworking and is then retested. No development occurs in the integrated testing environment. After development has been completed for all modules and they have been tested to the satisfaction of the architect, the user group is invited to participate in acceptance testing.

User acceptance testing is conducted in the final testing environment, which is configured to be nearly identical to the final production environment. Indeed, in some instances, the testing environment becomes the final production environment, whereas, in other instances, a successfully tested product is migrated into the existing production environment. User acceptance is conducted with users present and follows a script that tests each functionality as identified by the users in the original signed-off requirements. This is where a project manager needs to facilitate and mediate if users deviate from the principle that acceptance implies fulfillment of the project scope (no more and no less). Sometimes, users request enhancements at this stage and balk at signing off on acceptance. The SDLC approach is useful in explaining that enhancements are possible but should be held for a second iteration of the SDLC, resulting in a new version or release of the software product.

5.5 Governance

As part of any IT project, a project manager should recommend that a governance group be chartered to persist in life beyond the project and for the duration of the economic life cycle of the product. A governance group is typically composed of people from the user group, people from IT support, and people at a management level who can identify and seek sponsorship of enhancements and upgrades.

The charter for governance is to monitor the operational characteristics of the product, identify any gaps between desired and actual performance and capabilities, and conceptualize and trigger appropriate iterations of the SDLC related to the product. Governance is a legacy of any IT project. The project manager's role is to charter the group and ensure that its charter focuses on its sustainability for the duration of the economic life cycle of the product it is chartered to govern.

5.6 Training

Training involves more than teaching users the keystrokes necessary to use the system. When a new digital tool is introduced, or when significant changes are made to a tool and/or its related business process, it is important that users and any stakeholder affected be given the context and support to be able to adapt to those changes. The level and extent of training depends on the nature and extent of the changes.

There is an important opportunity during the initial planning, scoping, and IT tool selection processes to document the business process changes and the drivers for the change. This can be used throughout the project to keep perspective, and is especially useful during training. Training materials should include clear direction on roles and responsibilities for the system's use, any modifications to existing business processes, input and outputs, as well as the specifics of using the new system. In addition, by creating scenarios for testing and having users conduct the testing with those scenarios, they are receiving additional training that can be shared with others.

When a commercial software is being implemented, it often comes with generic training materials provided by the vendor. It is worth the additional effort to enhance this with utility- and process-specific context. Depending on the number of users affected by the system, it can be effective to define the "power users." These are staff designated to learn the system sooner and at a greater depth than most staff. Power users can be key players in acceptance testing, as trainers themselves, and as support for staff after the vendor and consultants have left and questions arise.

5.7 Upgrades and Enhancements

Software inevitably requires upgrades and enhancements. As the project progresses to releasing functionality, it is important to designate the team of people, typically a subset of the original implementation team, to manage, test, and implement future upgrades. During the initial project, the project manager's role with respect to upgrades and enhancements is to capture and document desirable ones as they are mooted by users and hand them over to the governance group upon project completion for appropriate action.

It is considered poor project management practice to incorporate scope changes without sponsor approval. Upgrades and enhancements constitute projects in their own right and can consume resources well beyond the approved funding levels for the base project. Many IT project managers have followed the siren song of upgrades and enhancements only to crash on the rocks of scope violation and budget overrun.

5.8 Version Control

Version control is a subset of data management in the PMO. Data management requires that project records be captured, archived, and retrieved. Many data records are developed iteratively or become superseded. Any effort of the project team that is based on superseded data or information is wasted effort and will require reworking, plus associated cost and time effects. Even worse, any effort based on superseded data or information that goes undetected poses significant risk and quality management issues to the PMO. Therefore, it is essential that rigorous version control be applied to all data and information.

Version control should be embedded in the documentation for software code, for technical memos, for decision documents, and so forth. In fact, version control should be applied to all products within a project so that no ambiguity will exist as to what version the reader or user is focusing on when data or information are retrieved.

6.0 PROGRAM MANAGEMENT: A PORTFOLIO OF PROJECTS

6.1 Organizational Aspects

Many organizations use standard project management practices to increase the likelihood of success for their technology initiatives. Often, however, the same organizations do not routinely undergo the process of project portfolio management (PPM). This is a methodology for analyzing and collectively managing a group of current or proposed projects based on a collection of key characteristics ("Project portfolio management," n.d.). It is a tool that organizations use to align their IT projects and initiatives with business goals; it optimizes their collective value and measures the performance of the individual projects and the collective program of projects and initiatives. In addition, use of PPM also provides indirect benefits as a result of standardization of processes, improved resource allocation, and increased opportunity for process improvement.

6.2 Methodology

There are many tools and methodologies available to assist organizations with PPM. The key challenge is gaining sponsorship for organizational commitment to start and maintain a PPM process. Project portfolio management has two distinct components: effective governance and evaluation.

6.2.1 Governance

Good governance of PPM requires a committee consisting of representatives from different departments, typically middle managers, who are familiar with the company's overall core business and short- and long-term business objectives and who understand senior management's mission and vision. Members of the committee are required to have strategic perspective and to use the company's overriding goals and objectives rather than those of their departments in making their governance decisions and trade-offs. The governance committee should develop and maintain a list of criteria based on the organizational business needs and then use these criteria to review the merit of each project individually and in combination with other projects. The process of developing the set of evaluation criteria is of vital importance for guiding the PPM process. Establishing evaluation criteria and supporting contextual documentation must be completed before the evaluation of projects begins. Some common evaluation criteria include the following (Baschab & Piot, 2003):

- *strategic value*—how the project will give the company new capabilities to have a positive effect, both internally and on the company's customers and suppliers.
- *financial value*—This includes the project cost and its return on investment (ROI). The ROI should consider both tangible benefits (e.g., cost reduction) and intangible benefits (e.g., improved flexibility).
- *adequacy of existing system*—If an IT project is intended to upgrade an existing system, the committee must consider the importance and adequacy of that system. Is the upgrade urgent? Could it be deferred for a year?
- *risk*—The committee should consider the risks associated with undertaking the project and the risks associated with canceling or deferring the project.
- *interdependency with other projects*—The sequencing of a particular project may be affected by schedule and status of other projects in the portfolio.

The aforementioned represent sample evaluation criteria. Organizations may include other criteria based on their type of work. For example, a public organization may consider the effect of projects on improving public safety, water quality, or community relationships.

6.2.2 Evaluation

Evaluation implies developing and implementing standard and repeatable processes for evaluating, approving, and sequencing projects and initiatives.

The governance committee evaluates all projects and initiatives considered by the organization on a periodical basis. Some organizations convene their committees on a monthly or quarterly basis or to coincide with preparation of an annual budget. For a project to be considered as a part of the portfolio, the requester must submit a project proposal. The project proposal is often in the form of a fill-in-the-blanks template that, at a minimum, includes the following:

- project name;
- project description: a short description of the project;

- project data: estimated cost, schedule, required resources, and risks;
- business case for the project: describing the business benefits of the project, that is, why the project is needed and what the expected business value is;
- financial metrics: ROI and net present value;
- relationship to other projects, that is, does the project complement or have a dependency on other projects that are either planned or already underway;
- risk: effect of deferring or canceling the project; and
- alternatives: the requester must provide at least one alternative.

The committee then reviews each project based on the criteria developed earlier. Several tools are available to assist in this process. The tools use a combination of spreadsheet tables and charts as a method of ranking projects.

The committee must discuss each proposed project. The chart in Figure 6.4 shows several projects ranked based on their strategic importance

FIGURE 6.4 Project Evaluation Process—Strategic Value, Project Risk, and Cost

and the quality of the existing system. In addition, the circle size indicates the financial commitment for the project (i.e., cost) and could be colored to indicate its financial performance (i.e., ROI). The committee must decide, for example, if the enterprise resource planning (ERP)/finance replacement project should proceed, which would require a commitment of most of the company's resources. The committee may consider if there is a subset of other projects that total a similar cost as the customer relationship management (CRM). Each project in the subset may rank lower than CRM but, collectively, may have more benefits to the organization. In other words, the committee must decide on the set of projects that will produce optimum benefit to the organization.

In addition to determining the optimum set of projects for the organization, the group must also determine sequencing of the projects. This depends, in part, on demand management and capacity management. The IT department may not have the available resources or skills to successfully deliver all the proposed projects and, thus, should be involved in sequencing of the projects. One of the most common reasons for project and portfolio failure is starting too many projects. Indeed, if the IT department takes on more projects than it has capacity to deliver, this could lead to undesirable organizational consequences.

7.0 PROJECT MANAGEMENT CHECKLIST

The list of management areas mentioned in this section may be useful for project managers handling IT projects. The PMI *PMBOK Guide* cited previously in this chapter recognizes and discusses many of these areas of project management. Organizing a project's document management and file structure around this checklist can prompt project personnel to pay attention to all of these areas when some areas might otherwise be overlooked. The project management checklist is as follows:

- *integration management*—a PMO activity centered on coordination,
- *time management*—schedule control,
- *scope management*—control and containment of project goals,
- *money management*—budget control,
- *quality management*—quality assurance and quality control,
- *resources management*—dealing with human resource issues,
- *communications management*—establishing how information is shared,

- *change management*—controlling scope evolution and time/money effects,
- *risk management*—anticipating/avoiding project failure,
- *data management*—capture of information/intellectual content,
- *procurement management*—purchasing and contract issues, and
- *standards management*—ensuring uniformity of design approaches.

The Project Management Institute has also developed a library of global standards under the themes of "projects," "programs," "people," "organizations," and "profession." According to PMI, the themes reflect the expansive nature of the project management profession. These standards are available for review on the company's website (www.pmi.org).

8.0 REFERENCES

Baschab, J., & Piot, J. (2003). *The executive's guide to information technology.* Wiley & Sons.

Project Management Institute. (2021). *Project Management Body of Knowledge (PMBOK) Guide* (7th ed.).

Project portfolio management. (n.d.). In *Wikipedia*. Retrieved March 2009 from http://en.wikipedia.org/wiki/Project_portfolio_management

Tuckman, B. (1965). Developmental sequence in small groups. *Psychological Bulletin*, *63*(6), 384–399.

7

Understanding Information Technology Processes and Practices

1.0 INTRODUCTION

Information technology (IT) represents a significant investment in almost any facility; In the mid-1990s, some estimates (Standish Group International, Inc., 1995) put the annual expenditure for IT projects in the United States at $250 billion. Out of those projects, 31.1% are canceled before completion, 52.7% end up costing at least 189% of their original estimate, and only 16.2% are completed within budget and on schedule. User requirements are frequently mentioned as one of the top five reasons for project failure, and 13% of project failures are attributed to inadequate user requirements (Standish Group International, 1995). Similarly, a Klynveld Peat Marwick Goerdeler Consulting survey cites that 61% of IT projects end in failure (Whittaker, 1997).

In 2021, the estimate for the United States spending on technology products, services, and staff was $1.94 trillion, and by 2022 the estimate will exceed $2 trillion (Statista, 2021).

Water and wastewater utilities manage infrastructure that includes facilities and pipelines; the scope and value of this infrastructure is large, especially in large urban areas. As part of their traditional mission, water and wastewater utilities have many years of experience with planning and executing standard (i.e., "brick-and-mortar") projects for designing, constructing, operating, or maintaining this infrastructure. Critical aspects of project management for these types of projects include scope, schedule, and budget.

The scope of traditional projects that deal with design and construction of physical facilities is typically defined and formalized by design documents. Two important characteristics of these design documents are that they are typically very detailed and that the design is fairly stable, meaning that the executed projects (with small and infrequent exceptions) closely follow the design documents. Furthermore, there is typically a contractually controlled change management procedure in place that defines such events as schedule changes, material changes, and similar deviations from the detailed specification.

By the time software projects came around, most utilities had developed a strong history and tradition of project management that was based on experiences and knowledge gained from previous years of managing standard engineering infrastructure projects. Therefore, there was a tendency to apply the same methodology to the design, development, configuration, and implementation of software projects. However, lessons learned from standard civil engineering design did not translate well to software development, acquisition, and implementation.

This chapter considers modern IT system technologies, from requirements, business process engineering, and systems architecture to software development and software procurement. It illustrates the evolution and diversity of frameworks, methodologies, and practices, and creates a foundation for undertaking IT projects.

2.0 ALIGNING BUSINESS INFORMATION TECHNOLOGY PERSPECTIVES WITH FACILITY OPERATIONAL TECHNOLOGY—INFORMATION TECHNOLOGY AND OPERATIONAL TECHNOLOGY

Most people understand the difference between traditional, enterprise IT systems and facility networks that are used to operate control systems. The business enterprise is typically referred to as simply the IT system—it is where people have access to email, software applications like Microsoft Word, Microsoft Excel, accounting systems used to run businesses, applications used to enter time for hours worked, and so on. The operational technology system, or operational technology system, is used for facility floor controls—whether someone is controlling an industrial facility making widgets or a water facility making clean water for the public. They are both networks, but very different in many ways.

2.1 Confidentiality, Integrity, and Availability Triad

For years the IT world has used the CIA triad to describe a model to follow. It is depicted as a triangle with each side supporting the next (see Figure 7.1). The acronym stands for confidentiality, integrity, and availability.

Confidentiality is of upmost importance in an IT network. Employees use the enterprise network every day to perform the many tasks associated with each business. Enterprise systems include data from human resource departments that may include all sorts of personal information associated with the businesses employees. Senior executives use the enterprise to store confidential data related to the state of the companies they run, data they do not want their competition to steal. The list goes on and on regarding the importance of confidentiality in any enterprise system, from military emails to hospital data that include patients' personal data—confidentiality cannot be compromised.

Integrity is the next word in the triad that describes the importance to have accurate and reliable data. There are many reasons data can become corrupted on an enterprise system, and these types of systems must make provision to securely store data for future retrieval, and to have systems in place to restore data in case of system corruption. Company data for inventory, for example, are important to the bottom line and to ordering and reordering; therefore, data must be accurate at all times to make good decisions.

Finally, availability refers to the network itself being available for users to have access to their data and software programs they use. Availability is not the most important for IT systems compared to the other two in the triad, but obviously still important. An IT system typically has scheduled

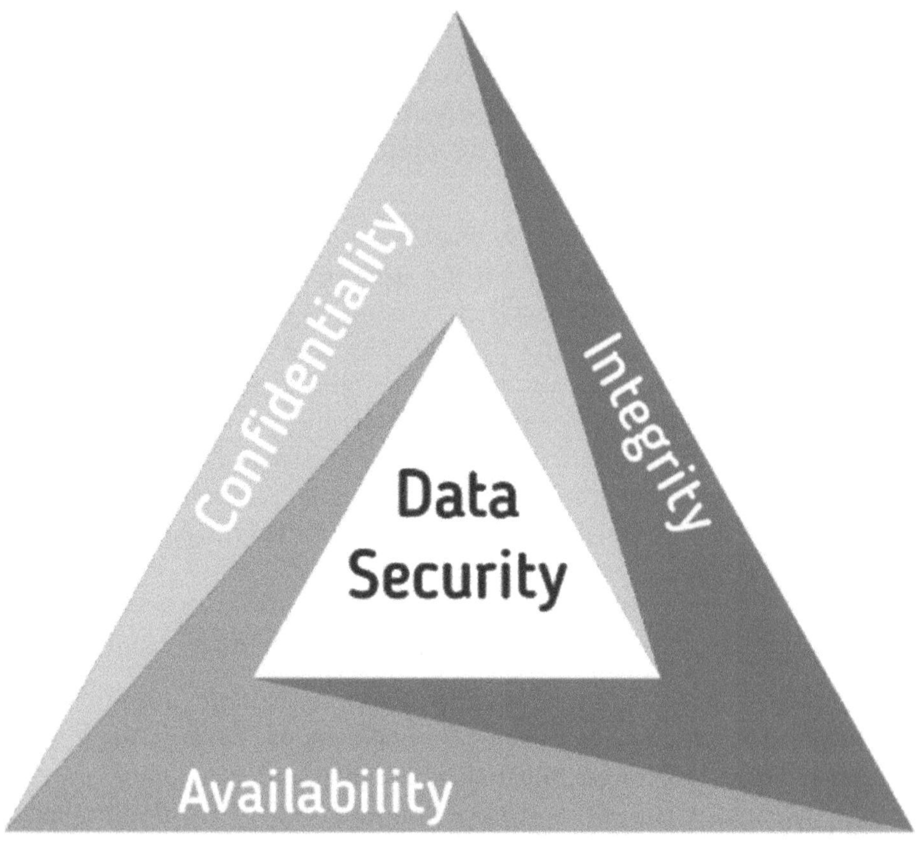

FIGURE 7.1 The CIA Information Security Triad

Note. Developedia from *Information Security Principles,* by Developedia, 2020 (https://devopedia.org/information-security-principles). CC BY-SA 3.0.

downtime for system maintenance, and this is generally done after business hours so as not to affect the productivity for the majority of the employees. So have some compassion on the IT staff!

For operational technology systems, this triad is turned upside down, or backward, to identify the model that these types of networks adhere to. The most important thing for operational technology systems is availability. These systems must continue to run or they will lose money. Even worse, if a water facility network goes down and is unable to make clean water for its customers, it can be catastrophic in many cases and cause all kinds of problems in the community.

2.2 Information Technology/Operational Technology Convergence

One of the common buzzwords in today's vernacular is IT/operational technology convergence. This represents an alignment of the IT department and their responsibilities and the staff who support the operational technology side of the network. For many years, these two entities have been at odds with one another in many facilities. The IT department is used to setting up networks, providing computers to employees, and protecting the company from loss of data, cyberattacks, and so forth. The staff on the operational technology side typically do not want the assistance of the IT departments because "they don't understand the needs of our system." Hopefully these days are quickly coming to pass where everyone is working together to create cybersecure systems, helping one another out. This must be the new philosophy for any enterprise to be successful in business. There is no alternative for success; teamwork and collaboration will make a huge difference when it comes to the bottom line.

2.3 Alignment

At this point, the old adage, "Easier said than done," can be heard. However, "Anything worth having will always require a good amount of effort." But in the end, the effort will pay off. It is important to have the leaders of IT and operational technology in constant communication, so everyone understands the business needs from the top down. It can be very beneficial to use the triad to help each group understand their priorities and why they are important. If technical staff can understand the "why," it will help in supporting everything else and it will open lines of communication that may prove to provide positive influence to make improvements in processes and workflows.

If these two groups can work together and understand how each system must operate, it can result in sharing resources and ultimately saving costs in the long run through common hardware, better support, and a better ability to serve the end user.

3.0 INFORMATION TECHNOLOGY ARCHITECTURE BASICS

For an IT network to function, it must have a mix of hardware and software, all configured properly. For an IT network to function safely, securely, and with resilience, it requires quite a bit of hardware and expertise to set it up and configure it well and right. Remember when "networking" systems first started? They were very simplistic—one unmanaged network hub. Things have changed quite a bit.

The OSI (open systems interconnection) model describes the seven layers that computer systems use when communicating over a network. Each layer has a different function in the communication scheme, and hardware devices are often referred to in their respective layers to define how they function. See Figure 7.2.

In the operational technology space, we often use the Purdue model to define our network segmentation. Segmentation will be discussed below in more detail.

3.1 Hardware

There are many hardware components that are associated with IT networks. This chapter will touch on some of the important ones; however, for more information refer to other chapters, especially Chapter 8, which discusses cybersecurity.

When discussing networks and hardware, it is helpful to distinguish the types of devices in terms of the layer in the OSI model they reside on. The first layer, Layer 1, is called the physical layer and refers to hardware that connects systems together. An example might be an Ethernet segment or

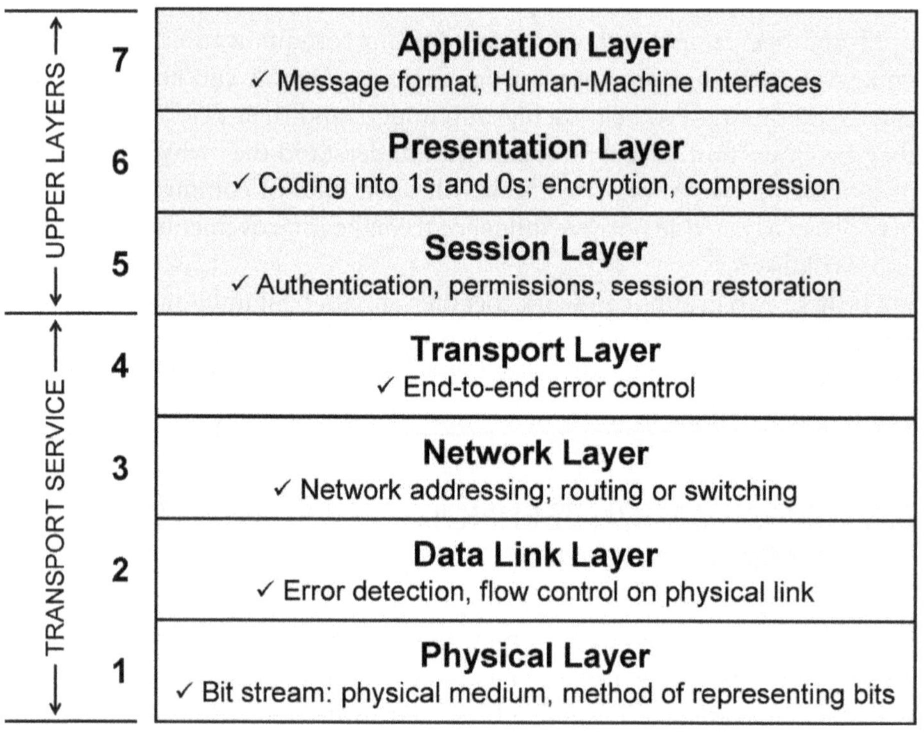

FIGURE 7.2 OSI Model

Note. Joe Manna Blog from *Here's How I Approach WordPress Troubleshooting*, by Joe Manna, 2014 (https://blog.joemanna.com/advice-troubleshoot-wordpress/). CC BY-SA 3.0.

serial connection. A Layer 2 device is a device at the data-link layer, defining how data are formatted and how connections are accomplished. An example might be a NIC (network interface card) or a simple/basic network switch. Layer 3 is referred to as the network layer and is typically composed of devices that are more flexible than the Layer 2 devices and can be configured in various ways. An example would be a router that is used to direct traffic on a WAN (wide area network). A Layer 3 switch is a hardware device that can be configured as a router.

In the IT world, we want to connect users to applications and provide a data path for communication across the enterprise. In the past (and sometimes still today), we typically use a Category 5 or Category 6 cable connected to a wall jack from our computer NIC. From the wall jack, there is a cable that goes back to a server room patch panel and finally to a network switch that will connect everyone to a common network. The more common way to connect computers or devices to the network is over wireless routers that hide in the ceiling. Installation is much easier and cleaner.

Switches support high data throughput and can be enabled to create subnets, VLANs (virtual local area networks), and are used for LAN (local area network) applications. A LAN is typically set up for use in a home or office environment, as opposed to a WAN (wide area network), which is typically supported through the use of routers.

Firewalls are common hardware devices used at locations that are exposed to outward-facing systems. These devices are able to be configured and locked down to allow traffic that is trusted through, and anything else is blocked. This can be helpful when keeping the bad guys out; however, it can also keep the good guys out when not configured correctly—but don't worry about that, they will always let it be known!

As the risk of cyberattacks increases, the availability of new hardware increases as well. New companies are constantly emerging along with new hardware and software technologies. One of these hardware devices is the data diode. This is a hardware device that acts just like a regular diode, but in the computer world. The data diode controls the flow of data in one direction only, which ultimately protects a network from a hacker getting access onto a network through the device. Because data can only flow in one direction, there is no possibility to access the network from the outside.

3.2 Software

Software related to IT systems is used to protect networks, monitor activity, configure hardware, or be used in applications. The focus here is strictly related to the IT network. Home and office networks often use similar software applications to protect from malware events. Malware, or malicious software, is used by hackers to disrupt network operation. Programs such as McAfee or Norton have been used for many years to defend against this type of attack.

Some of the more dangerous and prominent attacks on our IT systems today is called ransomware. Hackers infect computer systems typically through embedded links in emails or malicious websites. A ransomware attack must be identified immediately, or files and data can be locked and held until a ransom is paid, typically through a cryptocurrency. Once the ransom is paid, sometimes hackers will provide a key to unlock data files, but this can take hours, days, or even weeks to unencrypt files, sometimes not working at all. The best defense against ransomware attacks is education and training staff how to recognize a potential threat. If an email with a document attached is received that does not look right or was not expected, it is best to not open the document and call the individual before opening. Most ransomware attacks can be identified if people know what to look for. Misspelled words in the body, a generic reference to the recipient or from the person sending, a strange subject, and so on. There are software tools to help mitigate ransomware attacks/damage, but the best defense is training. Knowledge is power.

For IT professionals or those maintaining the IT networks, many tools are available to assist with different tasks. Network monitoring software can be useful to troubleshoot or detect throughput issues.

All IT networks should have some sort of threat detection/prevention software/hardware. Some of the best software available on the market today to protect network is advanced threat detection/protection software. This is software that uses artificial intelligence algorithms to identify behavior patterns and alert staff before the threat is able to cause major damage.

3.3 Networks

There are several types of networks in the IT and operational technology space. A typical home or office network is considered a LAN, or local area network. As the acronym indicates, these types of networks run locally and connect printers and computers to a common backbone. Computers, printers, and other devices can be physically connected through copper cables such as CAT 6 (Category 6) or connected wirelessly through Wi-Fi hardware, which is becoming the most common platform.

When connecting or sharing data over a greater distance—not just a single location—we use the term WAN, or wide area network. The internet is the largest WAN using many different technologies to connect everything together. Communication takes place over routers, Layer 3 switches, cellular modems, and even satellites.

It is helpful to think of networks like a spinal cord that connects everything together so communication can take place. In an operational technology system, if efficient and reliable communication can be established between all of the controllers and computers, the rest is easy—everything in the facility can be controlled manually, at the very least, through the control system. But if things are not communicating, information will not be available to make the choices needed to control the devices in a facility. This is why it is critical to get these networks operational as quickly as possible.

3.4 Redundancy

There has been a lot of wasted money and effort over the years when it comes to redundancy. Another old adage can be used here—"Just because you can, does not mean you should." When deciding where to implement redundancy, it is extremely important to know what is extremely important. Networks/communication systems are important, and implementing redundant systems for specific IT networks is money well spent. But do some research first to ensure that there is not a better, easier, more efficient way to accomplish the goal. If a specific network segment goes down, what are the repercussions?

3.5 Data

Data have been around forever, but when we are looking at IT systems, data take on a very important role. Data are some of the most important assets in a water facility, industrial facility, and enterprise network. One of the more popular buzzwords in today's vernacular is "big data." This is a result of our ability to gather all kinds of data and store that data easily and inexpensively. One thing that continually occurs is the cost of hardware going down. For example, in the 1980s, if someone had a 40 meg hard drive, they were the geek of the neighborhood. And when they purchased another 40 meg hard drive for a couple hundred dollars, they felt really good about all that power to store all that data—they will never run out of disk space! As time goes on, hardware has become less expensive and more expansive. Now it is possible to have a 1 TB drive (1 terabyte, or 1 million megabytes) on a personal laptop. As a result, people store all kinds of data because it is so inexpensive. Data are so important and easy to store that we now have what is called metadata—data about data! Enter big data. The key to overcome this excessive norm is to utilize different applications to manage different types of data on enterprise systems. Some data do not need to be stored for long; other data must be saved for years. Determine the differences, and set up IT systems to manage that data better.

3.6 Internet of Things

With competition comes innovation, which can be a good thing. Entrepreneurs want to develop products that are cutting edge and can make a positive effect in society. A new technology in the market, related to smart instrumentation and smart devices, is referred to as the IoT, or Internet of Things. In the industrial world where it is important to design hardware to survive in harsh environments, it is referred to as the IIoT, or the Industrial Internet of Things. This can refer to any device that is on the internet (has an IP address), and is made accessible on some LAN, potentially sharing data across the globe if necessary. Nowadays, through the advancement of technology and IoT, it is actually possible to look inside the refrigerator and see what is in there from anywhere in the world. Software can indicate if the milk is sour or the temperature is not correct. This can lead to big data, so it is important to be careful about what data are being gathered and stored. Although some of this can be fun, there is a good practical side to the IIoT. It is now possible to gather instrument data from distant locations over cellular or radio technologies without running any cable.

This technology has its application and can be useful in many cases. But with all great technologies comes risk, so it is important to make sure that networks are set up correctly so that network vulnerabilities are not created through one of these devices. Always use trusted sources for instrumentation and network devices.

3.7 Network Segmentation

Network segmentation is one of the most important concepts when it comes to creating cybersecure networks. This topic will be covered in detail in Chapter 8; however, we will cover this here at a high level. Network segmentation is done through configuration of the network Layer 3 switches. Segmenting a network is like protecting a ship on the sea, making it more resilient to failure or sinking. The hull of a ship is "segmented" into several smaller compartments so if one area of the ship's hull is damaged by an underwater iceberg, that smaller compartment may be compromised and fill with water, but the whole ship will not sink. In fact, all the other compartments are perfectly dry except the one where the hull was damaged. Network segmentation works the same way. Information technology personnel must configure the network into segments, so everything is not on the same network, but on smaller subnetworks. Communication can still flow if necessary and configured to do so, but if a hacker successfully breaks into one of the segments, they can only cause trouble to the few devices/users/data associated with that segment. Hopefully the advanced threat detection/

protection software will provide alerts so IT personnel can completely isolate the segment and shut the hacker down before major damage is done.

By proper use of network segmentation, the attack surface will be reduced and will mitigate the damage that can be caused by cybercriminals.

3.8 Resilience

Resilience is something that should be built into almost any IT/operational technology system. Resilience refers to a system's ability to continue to provide the functionality to the users who rely on them even under adverse conditions. Resilience is not something that just occurs in a system; it must be built into them. Information technology networks are critical systems that must keep running. For example, if we rely on our networks to keep running so that our staff can keep working when power goes out because of a lightning storm or failed transformer, we must build in a battery backup system or have the ability to connect generator power quickly and efficiently. That is building in resilience to a network. Or creating a fiber-optic loop as the network backbone so if a fiber fails or is cut, the system still communicates. Resilience must be well thought out and implemented so when the time comes, systems will not fail. This is critical and requires staff who are skilled and knowledgeable to assist in the design and implementation.

One word of caution when designing resilience into systems. Information technology networks can become complicated—sometimes so complicated that end users cannot maintain them. This should be avoided if possible. Systems should always be designed with the end user in mind.

4.0 OWNERSHIP, MANAGEMENT, AND CONTROL

Information technology networks are complex systems that require constant maintenance and oversight. If these systems are not managed properly or maintained, they have the potential to become a severe liability. Proper planning and forethought can mitigate this concern and provide future improvements that can lead to cost savings and efficiencies.

4.1 Cloud Versus On-Premises

As the speed of the internet continues to improve through the use of fiber-optic cable and higher speed hardware devices, the ability to host IT services in the cloud has not only become a reality, but a preference in many cases. Typical IT systems in the past (and some still today) were hosted on-premises, relying on expensive high-powered hardware, software, and staffing to maintain. Backup systems need to be in place and maintained to store data in case of system crashes or local site issues. Server closets require extra cooling and space. Software systems require constant upgrades and patches—and the list goes on. Owning, managing, and maintaining these IT systems can be a time-consuming business. These systems also take up quite a bit of space and require energy/power to keep going 24/7/365. These traditional on-premises systems (also known as "on-premise") have their place and application in certain circles, but they are time/labor/money–intensive systems.

As technology continues to improve, we will always have alternatives to "the way we've always done it." Several large corporations have stepped up to the plate and now offer cloud services for many IT applications. Instead of owning all the expensive high-end server hardware and software required to run many of the applications required to run business today, now the heavy lifting can be done in the cloud and hosted by a cloud provider. This is also known as "platform as a service" or PaaS. Many companies offer this service and facilitate a number of different services. This does not eliminate the need for IT professionals in our facilities, but it can eliminate a lot of expensive equipment and lower maintenance costs quite a bit. Budgeting also can become a much easier task with PaaS. Instead of owning and maintaining a lot of expensive hardware, the burden is put on the cloud providers. They will also provide all the software required (patching and maintaining it as well) and the cybersecurity services. Because they are responsible for many different clients' data, most have done a very good job with their cybersecurity posture.

The cloud is not for everyone or every business. Like anything else, it requires much forethought and planning to determine if each unique situation lends itself one way or the other.

4.2 Maintenance

Information technology systems, because of their importance and constant use, require constant and ongoing maintenance. Typically, maintenance activities can be completed on a schedule, to reduce system downtime and mitigate loss of productivity. Scheduling maintenance activities late at night or on weekends when usage is low is good practice and saves money and aggravation if things do not go as smoothly as anticipated. Expect the worst and hope for the best when making system upgrades.

It is also good practice to do beta testing of network upgrades to ensure that things will work how they are expected to work. This is good practice for operational technology networks as well. Owning some equipment used to set up a mock system or even develop a digital twin in a virtual environment can save a lot of time and money, as well as headaches too. The more employees using the system, the higher the probability that something will not work as intended. This is why extensive testing is good practice to ensure compatibility across software platforms.

Patching software is critically important to maintain a good cybersecurity posture. Hackers and bad actors like to exploit software vulnerabilities to infiltrate systems. The only way to mitigate these potential (and very real) risks is to have a good maintenance program with a focus on keeping up with all software patches. This is not easy to do and will require excellent organizational skills and planning, but once a good system is in place, it will not take much time or money to keep this going.

4.3 Risk Considerations

The National Institute of Standards and Technology (NIST) has come up with some excellent standards and tools to use or consider implementing to maintain a healthy IT enterprise network. The NIST 800 set of standards covers a wide spectrum of networking and cybersecurity concerns and provides the right tools for every enterprise. One of these tools is the Risk Management Framework (RMF) process. This is a process that a municipality or company can walk through to evaluate their unique risks associated with each system and determine the best path forward for evaluating, implementing, testing, and maintaining the enterprise system. This process requires IT professionals to evaluate all the potential risks and determine a response for each risk.

In the IT space, risk identification is the first step in achieving a cybersecure system, but it is also a step in the right direction in the development of a resilient system. If we can reduce our risk, we can achieve a more resilient IT enterprise. Once all risks have been identified, each one must be evaluated by the team responsible for the success of the business—whether it is the business of providing clean water to the public or making widgets in a factory. This team approach is the only way to properly accomplish the goals for the business, and it requires leaders from different backgrounds and perspectives to determine the tolerance.

Here are some potential IT risks:

- vendor panel hardware/software
- remote connections into the enterprise
- foreign device connections (e.g., cell phones over Wi-Fi or BYOD "bring your own device")
- access to physical servers/hardware

- lack of data backup, or poor storage procedures
- cybersecurity policies
- network hardware policies and procedures
- privilege or system accessibility
- no disaster recovery plan, or a poor plan

Not every person is necessarily involved in each risk factor to determine the tolerance, but everyone must provide input and feedback to areas of the business that are relevant to their area of expertise and responsibility. Each IT professional responsible for a system must identify their risk tolerance for each risk identified.

With limitations on funding and availability of workforce for most business enterprises, evaluating risks is an important task but needs to be prioritized. Using the NIST 800 RMF process helps to identify risks, evaluate the level of tolerance, and prioritize each mitigation path, addressing each risk in a timely manner. By considering and addressing all system risks, we can improve our resilience and provide better services to our customers.

5.0 DEFINING REQUIREMENTS

5.1 Hardware

Enterprise IT relies heavily on specific hardware devices to provide the physical backbone and infrastructure required to allow staff to communicate and run software packages to complete tasks. When it comes to hardware, it is imperative to use devices that are readily available and are trusted in the applications for which they are used. It is good practice to establish hardware standards for IT networks and not deviate unless there is a unique manufacturer requirement or technology improvements. By using a single vendor to provide network devices, it is possible to set up a predetermined pricing structure and/or obtain discounts through volume. In addition to pricing, when a single vendor is used for an enterprise and network issues begin to occur, a single vendor is more apt to assist in solving the problem and typically can provide the expertise necessary to identify and resolve problems. When hardware is mixed and the system is comprised of disparate equipment, it can cause issues that would not otherwise have occurred. When it comes time to resolving issues and vendors are contacted for support, it oftentimes turns into the blame game. Standardizing hardware and memorializing this in a standards manual for reference by vendors and internal staff can be worth its weight in gold.

When establishing these standards, always ensure these devices are expandable for future growth and they meet throughput expectations. It is good practice to select hardware that provides 50% expansion or more. Using flexible hardware products can provide expandability when it is needed.

Once standards are established, it is important to have access to spare devices should something fail or get damaged by weather events or other unexpected surprises. Sometimes vendors will work with customers to maintain a list of spare parts in the event of a failure. Each facility must determine what is critical to have on hand and what hardware is not as important. Only the implementation and maintenance staff can determine how the system is set up and how it will be repaired when something goes wrong. In the IT space, it is not a matter of "if" something will go wrong, but a matter of "when" something will go wrong. Hardware will fail, and a plan must be prepared ahead of time to quickly resolve these issues when they occur.

5.2 Software

All IT enterprise systems must provide an environment in which staff can use multiple software packages while maximizing their efficiency. In addition to user-specified software requirements, the enterprise itself relies on many different software applications to operate smoothly, efficiently, and securely.

Some software is used by all staff, and some software is specific to the needs of the individual. The challenge for IT professionals is to determine if the need for a specific individual is real and if there are alternative products that are already being used by others. It is important to limit the amount of software being used by staff because this software must be licensed and maintained. Licensing is fairly simple at first. Software is purchased and installed for use, but there may be more to deal with in the future. Many vendors have annual license fees. If these fees are not kept current each year, they will stop working unless the license is renewed. This takes both time and money for each unique piece of software on the enterprise. The bigger issue is with maintenance. To properly "maintain" software, attention must be given to any new patches that may come out related to security or improvements. When hackers find a vulnerability with any software package, they share it on the dark web, and it becomes an immediate liability. Once software vendors recognize a potential vulnerability, they will provide a patch to eliminate the vulnerability to users whose licenses are up to date; therefore, the more unique software packages that are installed on a system, the harder it is to keep up with maintenance and license fees, and the process it takes to stay current.

On the enterprise side, there are several software packages that all users require, and these too must be kept current and constantly patched and updated with the most recent versions. Software used by the enterprise includes operating systems, malware software, office suite software, and so forth. All of this software requires IT infrastructure/hardware to support throughput requirements. Software packages are changing all the time as a result of company requirements to do business. A good example is during the COVID-19 pandemic that swept the globe in 2020: Companies were forced to require employees to work from home for purposes of social distancing. The use of software packages such as Microsoft Teams and Zoom has put a heavy burden on many companies without much forewarning. The companies that had developed resilient, high-performing networks for their staff fared well during this time; other companies had to quickly make major improvements to hardware and software components, oftentimes having to wait because of the immediate increase in demand for network products. For IT professionals and the companies they work for and support—always be prepared for something to happen, it is not if but when.

6.0 PROCURING AND FINANCING SOFTWARE AND HARDWARE

There can be many hurdles to jump when trying to acquire hardware and software for a municipality, business, or company. Every entity has its struggles to deal with. The best and most popular way to purchase hardware and software is to have a good budgeting process each year for the next fiscal year. There is a benefit for a company to own its products, but these days there are almost always recurring costs, even with hardware.

Some hardware devices, for example, may use a special firmware device that needs to be updated regularly. Some artificial intelligent devices have this challenge, but the manufacturer in some case may require an annual fee to support the ongoing advancements of that technology.

Some software devices use a method called "software as a service." These vendors simply provide a software package (sometimes hardware as well) and provide a service associated with the customer needs and charge an annual, or ongoing, fee. In the cloud example above, a vendor provides all the hardware and software needed to support data storage and manipulation or monitoring of that data. The end user pays an annual fee, and everyone is happy.

These methods make budgeting much simpler. For example, IT managers would no longer need to worry about servers and software to support their operations if they rely on cloud servers and services. All they would need to do is budget annually for the cost of the service. No maintenance and no surprises.

In operational technology systems, projects may use financing options to implement them; therefore, these systems are completely financed, including all the software and hardware. Traditional IT systems do not typically take on this financing burden; however, if financing must be done to implement a large immediate need, it will be important to quantify all the recurring costs associated with the software and hardware as described above to understand how it will be paid off within an appropriate time. Software and hardware technologies tend to advance and change quickly, and it would not be prudent to be paying for old devices when the latest and greatest technology becomes available to help achieve technology/system goals.

7.0 REFERENCES

Developedia. (2020, July 21). *Information security principles*. https://devopedia.org/information-security-principles

Manna, J. (2014, November 24). Here's how I approach WordPress troubleshooting. *Joe Manna Blog*. https://blog.joemanna.com/advice-troubleshoot-wordpress/

Standish Group International, Inc. (1995). *CHAOS*. https://www.csus.edu/indiv/r/rengstorffj/obe152-spring02/articles/standishchaos.pdf

Statista. (2021). *Business and government spending on information and business technology in the United States from 2012 to 2022*. https://www.statista.com/statistics/821769/us-spending-it-products-services-staff/

Whittaker, B. (1997). *What went wrong? Unsuccessful information technology projects*. KPMG Consulting. http://p3m.com.au/Reference/KPMG_1997.pdf

8.0 SUGGESTED READING

Ingildsen, P., & Olsson, G. (2016). *Smart water utilities: Complexity made simple*. IWA Press.

8

Keeping Your Systems Safe: What You Need to Know About Cybersecurity

1.0 INTRODUCTION

1.1 Overview of the Threat Environment

1.1.1 Cybercrime As a Growth Industry

Cybercrime is a growth industry. Daily there are reports of another ransomware attack and cyberattacks on critical infrastructure. Virtual attacks have largely supplanted physical conflict between adversarial nation-states, and to put it bluntly, cyberwarfare is here. The reasons for this shift in paradigm are obvious: minimal cost, practically no risk, and large rewards; these rewards can be money or social discord, or as ominously appearing, physical damage and disruption. All these end results are possible and attainable given the present state of technology and the abundance of soft targets.

Threat actors, that is, those adversaries, attackers, and hackers who are attempting to breach and compromise computers and computer networks, are becoming sophisticated in their means and methods of cyberattack. Many of these adversaries are sponsored by nations interested in sowing discord and division, as in the Russian-sponsored attacks on the U.S. election process. Others, such as China, are more interested in appropriating technology. Other nation-states such as Iran and North Korea are interested in creating mayhem and physical damage to infrastructure. Still others are specializing in ransomware attacks designed solely for monetary gain. All of these threat actors have found fertile ground in the West because of widespread freedom afforded its citizens and the relatively unregulated electronic frontier, specifically the internet.

Unfortunately, the West does not take the threat seriously, and a large portion of the populace do not have the knowledge or skills to recognize or guard against even the most rudimentary cyberattack. The purpose of this chapter is to provide some basic concepts and ideas to allow managers and operators to make informed decisions about cybersecurity and provide a defense to this burgeoning threat to national security.

1.1.2 Soft Targets

Typically, a soft target is a facility that has lax or nonexistent cybersecurity. Cybersecurity is a broad term describing the suite of policies, controls, and practices that are used to protect a facility. With the recent spate of ransomware attacks, many high-value targets are hardening, placing them out of the reach of threat actors. The natural response is to move on to lower hanging fruit—those facilities that are relatively unprotected. Soft targets typically do not enforce cybersecurity policies, or do not have any policies in place. Without any sort of policy in place, there can be no controls or approved practices to safeguard the facility. This lack of protection and the culture that it develops is the definition of a soft target.

An effective method of determining if a facility is a soft target is to do an audit of exiting cybersecurity measures in place, followed by a "penetration test" performed by a consultant that specializes in this type of testing. Penetration testing is discussed in some detail in Section 4 of this chapter; however, for this discussion, penetration testing is a series of tests done by "white hat" hackers who are typically certified ethical hackers or similarly trained software and hardware professionals. These tests simulate an actual cyberattack and can identify areas of vulnerability.

1.2 The Need for Critical Infrastructure Cybersecurity

Water and wastewater treatment facilities are critical infrastructure. Critical infrastructure is defined as "systems and assets, whether physical or virtual, so vital to the United States that the incapacity or destruction of such systems and assets would have a debilitating impact on cybersecurity, national economic security, national public health or safety, or any combination of those matters" (Exec. Order No. 13,636, 2013). As has been demonstrated by recent events, threat actors will seek out vulnerable public infrastructure to cause disruption and division and seek to damage a facility. Several treatment facilities have been targeted; Oldsmar, Florida, is the most recent example of a successful breach. This breach can be traced directly back to human error but was aided and abetted using an unsecure network and outdated computer equipment. A lack of cyberhygiene was a major contributing factor.

1.3 Cybersecurity and Cyberhygiene

The term *cyberhygiene* is used throughout this chapter to describe a culture of cyber awareness and best practice to avoid being compromised by threat actors. Cyberhygiene is a combination of policies, practices, and culture that enables a facility to standardize facility defenses across the several levels of staff capabilities; cyberhygiene can be taught but must also be enforced. It is up to the management and administration of a facility or authority to set and enforce these policies. In cybersecurity, the expression "An ounce of prevention is worth a pound of cure" has never been so apt.

Cyberhygiene also adds cost to capital and operating budgets. The Florida breach was the result of a lack of policy and by human error, but it was also the result of a failure to properly fund the technical infrastructure necessary for constructing and operating a modern and robust cyber defense based on the latest technology. The use of older computers and remote access software, along with poor password policy, was the perfect storm that led to a serious breach. It is somewhat fortunate that the facility's physical characteristics prevented serious damage to the process systems and the public health.

1.4 Chapter Summary

The purpose of this chapter is to educate the reader about the form and functions of cyberwarfare, and to provide methods of protecting critical facilities from a cyberattack. This chapter is not exhaustive; to present a complete manual on cybersecurity for critical infrastructure would dwarf this book. The material presented will provide a starting point and a basis upon which the reader may build to gain a more complete understanding of the "threat landscape" and methods of protection and mitigation.

This chapter is not a technical treatment of the subject. This chapter is written to allow nontechnical staff to access information that is considered vital to understanding the threats and the countermeasures required to design a defensive system. The chapter contains a detailed suggested reading list that points the reader to sources of in-depth and detailed information for further study.

Section 2, Means and Methods of Cyberattacks, will describe methods of cyberattack and provide some basic information on how attacks are perpetrated. It should be noted from the start that the human asset is the most easily and most often compromised element of any system. This has much to do with education and what can be considered a "sense of ownership" in the facility.

Section 3, Social Engineering, describes what is perhaps the most pernicious manifestation of cyberwarfare. Building on the concept of the human asset as an integral part of the system, this section describes how the human asset is manipulated into making decisions that are injurious not only to themselves, but also to the organization they work for. Social media is described; social media is a major source of information used to build attack profiles on an individual or an organization and is pervasive. The human asset at all levels is susceptible to these manipulations and attacks, and more than 80% of successful breaches have started with a social engineering attack.

Section 4, Prevention Techniques, describes how a facility can build a robust defense by using proven techniques, education, technical improvements, and culture change. This section is a general discussion of the state of the art and should be used as a basis for further study or to inform before retaining specialized talent to evaluate a facility, train staff, or harden your defenses.

A detailed suggested reading list follows Section 4 to aid the reader in further research and study. Many of the books and publications are available online free of charge and some for a small fee. It is highly recommended that the reader starts their course of study here and then seeks out more detailed treatments of the areas of interest described here.

2.0 MEANS AND METHODS OF CYBERATTACKS

This section will discuss the various means and methods of a cyberattack. This information is important because it will allow managers and operators to understand the threat landscape and then take steps to protect themselves and their facilities from attacks. This is not a technical discussion; the information provided below is meant to illustrate in lay terms the methods of attack and the tools a threat actor uses to conduct and complete a successful cyberattack. The terms "threat actor," "hacker," and "attacker" are used interchangeably throughout.

Cybercrime and cyberwarfare are a fact of modern life. Nation-states, hostile or disgruntled employees, "hacktivists," and social engineers pose a clear and present danger to our nation's infrastructure and critical facilities. Cyberattacks are low-risk and high-payoff activities because of the abundance of soft targets in the cybersphere; as higher value targets begin to harden and reduce their online attack surface, threat actors will seek out and attack soft targets first. As will be seen, their arsenal is vast and sophisticated. Soft targets are those targets with exploitable vulnerabilities; most vulnerabilities can be identified and mitigated if an organization or facility has the will and the budget to do so. The alternative is a harmful breach of a supervisory control and data acquisition (SCADA) system or office network, corruption of servers or entire servers, or facilities held for ransom.

The chapter starts with a description of the attack sequence, called a "kill chain" and then describe the various tools in the attacker's arsenal. In Section 4, methods of protecting a facility from a breach will be discussed, and this section will also provide an understanding of how an attack is perpetrated.

2.1 The Cyber Kill Chain

Every cyberattack follows a plan, or chain of events designed around a profile developed for a prospective target. This is called a "kill chain," the term being borrowed from the military and describing the planning and execution of a cyberattack. The kill chain concept is used to illustrate the structure of an attack: identification of the target, dispatching the force needed to attack the target, the actual attack, and finally, the elimination of the target. The concept was refined and popularized by Lockheed Martin for use in cyberwarfare. The steps of the kill chain allow analysts to gain greater visibility into a cyberattack and achieve a greater understanding of a threat actor's means and methods of attack. The steps of the cyber kill chain model are as follows:

1. reconnaissance: accumulating information about a target, including email addresses, personal information, open-source information, and so forth
2. weaponization: developing an exploit based upon accumulated information into a deliverable payload
3. delivery: delivering the weaponized exploit via email, USB, infected files
4. exploitation: exploiting a vulnerability on the target to allow installation of the payload
5. installation: installing the payload on the target asset

6. command and control: activation of a control channel for the remote manipulation of the target
7. actions on objectives: attacker now has control of the target and can execute further exploits or attacks at will

This model allows an in-depth analysis of an attack and the ability to develop methods to defend against an attack. All cyberattacks follow some form of this model. Aside from providing a framework for understanding the attack sequence, it allows further analysis into the attacker's means and methods, allowing analysts to develop means of "breaking the chain." The defensive techniques developed because of this analysis also serve to reveal the vulnerabilities of a target system. Defensive exercises such as red/blue team and penetration testing both follow some kill chain model in testing a system's vulnerability. These techniques will be discussed in Section 4.

2.2 The Human Cyber Kill Chain

An adaptation of the above model is the human cyber kill chain, which specifically deals with the exploitation of a human asset. This model has only four steps:

1. Research and reconnaissance—"recon": This phase of the attack involves gathering information about a prospective target. This information is typically open-source intelligence (OSINT) and human intelligence (HUMINT) obtained through social media and interpersonal interaction. This information is used to develop a basic attack profile on the intended victim.
2. Foster relationship—"hook": This phase involves developing a relationship with the target, whether actual or virtual. This phase allows the attacker to further refine the attack profile by interacting with the target in either an overt or covert manner. Targets will reveal sensitive information to a trusted acquaintance or a prospective business partner or love interest. This step involves various confidence techniques.
3. Exploit—"exploit": This phase exploits the vulnerabilities discovered through the recon and hook phases of the attack. Exploits can be technical, social, or physical in execution, and can be part of a multipronged attack on a victim. Technical exploits are done through email or file transfers; social exploits are carried out through actual or virtual relationships with the target; physical exploits are perpetrated "on the ground" and involve theft, physical manipulation, or interpersonal deception.

4. Exit—"exit": This phase involves the departure of the attacker from the target venue. This phase also involves the exfiltration of data, installation of command-and-control programs or disruption of the target system. The goal is to accomplish a successful attack and ideally will leave no traceable evidence of the attack. Typically, the attacker will retain control of the target system while actively evading intrusion detection systems.

Understanding the stages of a cyberattack allows us to develop methods to avoid compromise and break the attack chain. In the following section, methods of attacks used to compromise a target will be discussed with descriptions of how they work. Section 4 will delve into the methods of detecting and deterring these attacks. It should be noted that a majority of attacks are a combination of technical and social attacks; physical attacks are targeted toward hard assets, such as documents and equipment, though physical attacks can be used to obtain access credentials and human intelligence that are otherwise unobtainable through technical means. Many attacks use elements of all three types, while some use only one method; this has much to do with the attacker's resources and the motivation for an attack.

The following sections will describe the various methods of cyberattacks and the means by which they are perpetrated. This list is not exhaustive, however, and the reader is advised to pursue further research and training to gain a complete understanding of the threat landscape and the techniques used to defend a facility from an attack.

2.3 Technical Attacks

Technical attacks rely on compromising the technical aspect and assets of an organization. This includes the following:

2.3.1 Malware

Malware is malicious software. Malware is designed to disrupt, damage, or allow unauthorized access to an infected computer. Malware comes in several forms, which will be described below. Malware is most often delivered in emails, though infected PDF files are known to contain malware that executes upon opening the file. Malware is the primary threat that exists online but is largely preventable by observing commonsense rules and by employing robust intruder detection and mitigation "endpoint" protection. Enforceable password policy and practiced cyberhygiene bolster defenses and will thwart an attacker, at least making it more difficult for an attack to be successfully perpetrated.

This section will describe the several variants of malware that are prevalent online. This is not an exhaustive discussion and should allow the reader to further research the subject, which is extensive and can be quite technical. It should be noted that more than 90% of malware is delivered via email; this is accomplished though phishing, vishing, and smishing, which will be described below. Malware is also delivered by social engineering, which uses interpersonal methods of compromise and will be discussed in Section 3.

2.3.2 Trojan Horses

A Trojan horse is a strain of malware that is embedded in what appears to be a legitimate file or program. A "Trojan" will reside within a system, sometimes actively hiding its presence and evading antivirus software. However, once the Trojan is activated, it can install all manner of malware, including those described below. The most common malware installed are reverse shell programs, which initiate a connection back to a "command-and-control" computer in another part of the world. After that is established, the threat actor can install any other malware at leisure or monitor the victim's activities. Another common malware installed is a keylogger, which records keyboard entry sequences such as usernames and passwords and send them back to the command and control operator.

2.3.3 Viruses and Worms

Perhaps the earliest known cyberattacks were conducted by researchers and hobbyists experimenting with code in their laboratories. Like a story from science fiction, these experiments escaped the lab and, in some cases, shut down huge swaths of the nascent internet.

At their most basic form, viruses, and their related subspecies, worms, are self-replicating computer code that is designed to cause disruption and damage to a targeted system. Viruses differ from worms in one important way: While viruses are designed to replicate indiscriminately throughout a system, worms are typically purpose built for targeted attacks. Another important difference is that a virus requires a host as in a biological virus, while a worm does not; a worm can self-replicate across a network without relying on specific files as would a virus. The important distinction is that a virus typically infects a file and then replicates across a system, while a worm will replicate freely using network resources without relying on a host file to transfer to another system.

2.3.4 Bots and Botnets

A bot, or robot, is a computer that has been remotely appropriated by an attacker. In other words, the computer has become a remotely controlled weapon that can be used to attack other computers and systems as a proxy for the threat actor controlling it. A botnet, or robot network, is a collection of remotely controlled computers that can be used as an electronic weapon of mass disruption; this is no joke—botnets comprising hundreds of computers have been successfully used to attack websites and take them down using the distributed denial of service (DDOS) attack method.

A bot is created when a victim downloads malware that creates a reverse shell; a reverse shell is a connection back to a command-and-control operator who can then instruct the compromised computer to do their bidding. In most cases, the owner of the compromised computer doesn't even know that their system has been compromised and is perpetrating crimes. It is conceivable that botnets comprising thousands of computers are possible. As has been explained, the malware that makes this possible is typically downloaded when a victim clicks on a link or opens a file.

2.3.5 Spyware

Spyware is malware that resides in the background and simply records everything the victim does on the computer. This includes every keystroke, every website visited, every social media post, to name a few. The data gathered by spyware can be used for many purposes; passwords and other access credentials are the most obvious and common data that are appropriated and then either sold or used to compromise other systems. Spyware can be relatively benign; it can be part of an app associated with a retailer or other entity that is used for marketing purposes—this type of spyware is typically willingly installed by a consumer in exchange for perks like special sales or coupons—it is important to read the fine print of any app that is installed on an electronic device. It is common for something as presumably innocent as a weather app to require the user to agree to terms that stipulate the user give permission for the app to record or monitor all the user's activities.

2.3.6 Ransomware

Ransomware is a particularly dangerous strain of malware that has gained popularity in the last few years. There have been several high-profile instances of hospitals, transit authorities, and city government networks and particularly file servers being held for ransom. Ransomware is installed on a computer in the same manner as any other malware—by a victim opening an infected file or clicking on a link. Typically, the malware will not infect the victim's computer; rather, it will wait for the user to connect to a targeted network and then propagate throughout the network, looking for vulnerabilities. As the malware propagates, it infects other machines on the network. When the malware has compromised the targeted organization, and this is being monitored by the command-and-control operator, the malware executes and encrypts files on every infected machine and server. A message typically pops up stating that the organization has been attacked and asks for a payment in exchange for a key to unlock the files. There is no guarantee that the key will work after payment is made, however, and many organizations that have survived this sort of attack spend months rebuilding their servers and recovering from the attack.

2.3.7 Zero-Day Vulnerabilities

A "zero-day" vulnerability is a software exploit that is unknown to even the developers of the software. These vulnerabilities have no defense, and many successful attack campaigns have made great use of these vulnerabilities to effect costly and destructive attacks. These vulnerabilities are present in nearly every software package on the market. Many vulnerabilities are yet to be uncovered, and several entities, on both the black and white side of the issue, are actively searching for them; on the dark side, these vulnerabilities can be used for nefarious purpose or can be sold for large sums of money to nation-states or cyberterrorists. On the white hat side, these vulnerabilities are identified and patched by the responsible vendor. Many vendors, such as Microsoft, offer bounties for zero-day exploits. Zero-day vulnerabilities are one reason why systems must regularly be upgraded and patched to ensure that these systems are protected. This, it should be noted, is a supply side problem; these vulnerabilities are present in the finished product because of a lack of thorough testing and a rush to market. Fortunately, however, this is changing as the threat of zero-day vulnerabilities has been recognized.

2.4 Social Attacks

2.4.1 Phishing

Phishing is, by far, the most common social engineering attack regularly perpetrated by threat actors. Social engineering is a broad and complex (and controversial) topic that will be covered briefly in Section 3. Phishing gets its name from the "phone phreaks" who learned how to compromise the telephone system for free telephone calls or to control parts of the system for personal gain. The phone phreaks were the earliest hackers, and many of the techniques they developed to compromise the system were easily adapted for cyberattack—in fact, their methods provided the framework around which the modern hacking paradigm was created.

Phishing involves a targeted or indiscriminate attack using email as its primary means of delivery. More than 80% of successful cyberattacks start with a phishing email. The technique itself is simple: send the victim or victims an email asking them to click on a link or reply to the email—this, in itself constitutes the attack. The emails can seem quite authentic, and many people are tricked into responding, some with passwords and other access credentials, including but not limited to Social Security numbers, security question responses, online account numbers, and credit card numbers. Regardless of all attempts at education and interdiction, at enormous cost to an organization, people still fall victim to well-crafted phishing campaigns.

Phishing emails that are targeted at an organization will typically, as a first step, be sent to nearly everyone with the same email domain address (@xyzcorp.com); a well-administered email system will move these mass emails to the junk folder, though many systems need to be "trained" before they will properly identify the scam email. Phishing emails can come in many forms, but they also have some commonalities. For instance, spelling is a giveaway; look for misspellings in the name and body of the email, extra spaces, and special characters. A favorite technique is the substitution of "@" for "o" or "1" for "I," or vice versa. Many threat actors do not have an adequate command of the language or syntax of the target, and many are simply following a script without having much in-depth knowledge of the attack, its goals, or the methods used to affect the attack (these people are called "script kiddies," because they have little technical skill and simply follow a script).

Phishing relies on several human characteristics to be successful. It should be understood that phishing is a popular method of attack simply because it works; it is estimated that at least 10% of an organization's staff will still click on a well-designed phishing email despite all training or notifications on the threat. Phishing emails target human traits—need and greed. Furthermore, curiosity and general laziness or ineptitude provide fertile ground for these attacks to take hold and become successful exploits.

2.4.2 Spear Phishing

Spear phishing is a targeted email campaign designed around a specific asset that a threat actor has identified as a preferred target. This target can be a person or a facility, or an entire organization. Spear phishing relies on good intelligence and is typically preceded by an extensive intelligence-gathering campaign through the various means previously described in the cyber kill chain above. The targets are not necessarily "high value" but are situated in a position within the organization that, if compromised, would allow profitable access to the organization. These targets would possess valuable access credentials to systems or assets that can be sold or appropriated by the attacker.

2.4.3 Whaling

Whaling is a form of spear phishing that targets very high-value targets such as government officials, CEOs of a target organization, politicians, and other high-profile individuals, particularly celebrities. These attacks are often perpetrated by nation-states as part of a larger campaign of disruption or as a component of a multifaceted attack on an organization to obtain trade secrets or for ransomware attacks. Many of these attacks are not reported because of the embarrassment or exposure that would follow disclosure.

2.4.4 Smishing

Smishing is a relatively new type of attack that uses SMS, or text messaging, to perpetrate the attack. Like phishing emails, these messages offer goods or services that can be obtained by simply answering the text message or clicking on a link. Smishing uses many of the same deceptive techniques as conventional phishing, such as purposeful misspellings, appeals to need and greed, or offers of easy money or sex.

2.4.5 Vishing

Vishing is voice phishing. Typically, this attack takes the form of a telephone call from an individual purporting to be from a government agency (e.g., the IRS), or a delivery service (e.g., FedEx or UPS) or the ever-popular Microsoft Tech Support agent. Most of these entities do not call to pursue their purpose, so any call from them is suspect. The goal of the attacker is to make the victim believe they are legitimate and then trust them enough to hand over credentials such as Social Security and account numbers, with the required passwords. This approach is called "pretexting," and will be discussed below. This type of attack is on the wane, but it is still successful in compromising some socioeconomic groups, such as the elderly.

2.4.6 Distributed Denial of Service

Distributed denial of service attacks use massive floods of emails or website access requests to overwhelm a web server and make it unreachable or to cause it to crash. These attacks are still possible, but protocols put in place after several high-profile DDOS attacks have made this attack ineffective and unprofitable. Unprotected systems, or systems that have not been upgraded or patched, are still susceptible to this type of attack.

2.4.7 Pretexting

Pretexting is the art of creating a fake persona to trick a victim into trusting the attacker. Pretexting makes great use of open-source intelligence (OSINT) that is mined from social media and publicly available information about a victim. Pretexting is a form of direct attack that requires the attacker to be in contact with the victim and to develop a relationship so the victim can be manipulated into providing the targeted information or access to a targeted system. Pretexting can take many forms depending upon the reconnaissance gathered about a target: job offers, offers of companionship or sex, or something for nothing or for a small fee.

An example of pretexting would be a threat actor calling a family member purporting to be a medical professional or a policeman, telling the victim that a loved one was in an accident. The perpetrator would then offer up information that was found online but would be familiar to the victim. This fosters trust in the victim for the caller, who then states that the victim needs to wire money to a special account "that is secure and protected" so that their loved one can be properly treated "so their life can be saved." This scam has several variants, a popular one being that the loved one is being arrested, and the victim must send bail money to the specified account immediately. Another is an online romance where the attacker preys on a lonely woman or man to send them money so they can fly to a rendezvous, but somehow keeps getting into jams that only a fresh infusion of money can cure.

In the first two scenarios, the attacker knew the whereabouts of the "loved one" and the name and contact information of the intended victim. In the lonely-hearts example, the attacker mined information that was posted in online chat rooms or dating websites. In all cases, personal information is typically gleaned from social media sites where both the loved one and the victim posted what they thought was innocent information that was ultimately used against them. A popular commercial shows a woman excitedly posting to her social media account about her impending vacation—cut to the burglar, who promptly thanks her for the heads-up.

2.4.8 Reverse Social Engineering

Reverse social engineering is a form of social engineering where, instead of an attacker approaching the victim, the victim is tricked into contacting the attacker. This is typically done by introducing some sort of crisis, such as a malfunctioning computer or mobile phone. The attacker has likely created this problem through some form of attack, like a phishing email that has executed on the victim's machine. The attacker then advertises their services, which the victim thinks is legitimate and contacts them for assistance. This type of attack takes several forms, but all rely on creating some sort of need or problem and then manipulating the victim into reaching out for assistance.

2.4.9 Water Holing

Water holing is another form of social engineering attack that use websites that a victim regularly uses and thereby trusts. The victim may be savvy enough to avoid clicking on a phishing email but would feel comfortable clicking on a link on a website they regularly access. Threat actors will gather intelligence about an intended victim using the methods described above to determine the victim's browsing habits. They will then analyze the website for vulnerabilities and inject malware into it. If the victim clicks on that link from inside of a secure system, the attacker will gain access to that system. This is one technique used to breach supposedly secure systems and stems from the victim's lack of security awareness.

2.4.10 Baiting

Baiting plays to the "need and greed" impulse in many people. Baiting is useful for installing malware and can be considered "clickbait on steroids." Baiting also makes use of open-source intelligence gathering to determine likely targets. Baiting typically offers something for nothing or presents a desirable product at a reduced price. Free software, music, images, or spreadsheets, or pornography, sexual liaisons, even drugs, are used as bait for the unsuspecting. Clicking on a link for these offers installs malware that allows the attacker to gain control of the victim's computer and then possibly the victim's employer's network, which may be the ultimate target.

Another form of baiting is using USB drives, which is a particularly effective method of breaching air-gapped facilities. The technique is simple and straightforward: Infected USB drives are left in venues where employees of a targeted facility or organization congregate. The normal impulse is curiosity or the desire to return the drive to the owner. The victim plugs the USB drive into their computer, and malware is installed. The victim can then remove the drive or even dispose of it and not be aware that they have been compromised. When the victim connects to the target network, the malware executes, completing the attack.

2.4.11 Quid Pro Quo

Similar to baiting, this attack offers some benefit to the victim for providing information. Social media is rife with this type of scam; this adds to the value of social media as a primary means of intelligence gathering and attack vectors. Greed and indifference are the key human characteristics these attacks rely on. The attack is a simple exchange, typically not to the victim's benefit. A study done in Great Britain showed that about half of random people stopped in the subway would willingly give away their computer or network passwords in exchange for a piece of candy or a cheap pen, or some other trinket (BBC News, 2004). No other information was taken from these people, and the study was taken no further. The study illustrated the callous disregard many people have for proper cyberhygiene and has been cited many times as justification for cybersecurity education and stricter password policies.

2.5 Physical Attacks

2.5.1 Dumpster Diving

Dumpster diving is another legacy of the phone phreak era. It is essentially what it says: a person sifting through a dumpster (or any trash receptacle) searching for technical data, credentials, information on the roles and responsibilities of key employees, or lists of expired passwords that can be input into sophisticated algorithms used to guess, or "brute force" system passwords. After gaining physical access to a facility, the well-prepared attacker can access any employee's work area and their trash.

2.5.2 Shoulder Surfing

Shoulder surfing is a technique that predates computer technology. An attacker with specialized training, or with a natural gift for memory, can simply stand behind an employee and record their keystrokes or what is being put on their screens. Computer screens are easily seen from the person in the next seat on a plane or train. Phone passwords can be easily spied on while being punched in. This can also be done from off-site if the employee's workspace is near a window and using high-end optical equipment. Long-distance parabolic or "shotgun" microphones are also used to obtain audio of privileged conversations between key employees. These techniques have always been a staple of spy craft, and the only defense is awareness and diligence.

2.5.3 Theft

Theft of critical documents or software is also a staple of cyberwarfare and has been used throughout history to obtain information that can be used to good effect by an adversary. Critical information stored online—in the "cloud"—is presumably secure, but the same was thought of the personal information retained by department stores and mobile telephone providers. It should be remembered that whenever any data leave a facility's premises, the owner loses control of them, and the data may be compromised at some time. Attackers see this information as high value and fungible. Best practice is to keep critical data on premises or within a secure facility that is owned and operated by that organization.

2.5.4 Tailgating

Tailgating is a physical attack that also relies on pretexting, but in an interpersonal setting. It is a common method of gaining physical access to a facility to access information or systems. The attacker poses as a delivery person, replete with a uniform and carrying boxes; the attacker asks a helpful individual to hold the door because their hands are full—the individual, wanting to be helpful, happily obliges. A variant is the "I forgot my ID card" technique where the attacker poses as a legitimate employee who forgot their access card. The legitimate employee obliges, and the attacker then copies the card or sells it before it can be deleted. The simplest form is an attacker simply following a legitimate employee through an open door, observing all social protocols, and carrying a briefcase or some other convincing prop.

2.5.5 The Long and Short Con

Confidence games or "cons" go back to ancient Greece. The con is a scam to take advantage of the victim's trust and to exchange something, typically to the victim's disadvantage. The con is a simple form of human intelligence, or HUMINT, which will be discussed in the next section. The "long con" typically takes place over several days or weeks and typically involves pretexting to gain the victim's confidence and trust. It may be carried out by a single individual or by a team. The short con happens in a matter of minutes or seconds and is generally broad in nature. It can be as simple as asking someone to hold a door open to a secure area while posing as a deliveryman with their hands full—again, pretexting on a simple scale. Either con is well practiced, and the perpetrators are seasoned professionals. Most "con artists" rely on a victim's propensity for "need and greed" or their willingness to help. The entire confidence game relies on those human characteristics that are easy to exploit, such as social incompetence, naivete, or nefarious impulses.

2.5.6 Side Channel Attacks

Side channel attacks are a relatively new phenomenon involving the use of the physical characteristics of a computer or computer peripheral. It has been shown by researchers that by intercepting electromagnetic radiation, observing indicators, or monitoring sounds made by electronic systems, data can be extracted from the system without physical contact. Side channel attacks have been known for decades and forms of electromagnetic eavesdropping have been used effectively since World War II. All electronic equipment leaks electromagnetic radiation, and this radiation can be intercepted at a distance. More recently, it has been found that power indicators on speakers or USB hubs can be used to recover audio that can be translated into useful data—this is called a "Glowworm" attack. These types of attacks are grouped under an NSA specification called TEMPEST—Telecommunications Electronics Material Protected from Emanating Spurious Transmissions—and is a secret program to protect sensitive data from being intercepted. TEMPEST has been replaced with the acronym EMSEC—emissions security.

EMSEC applies to all information systems. Presently, many devices can emit various degrees of electromagnetic radiation into the atmosphere or into a nearby conductive medium, such as wiring or plumbing. These emissions contain information the device is displaying or storing, creating, or transmitting and can be monitored at different ranges depending upon the device. Some are more susceptible to eavesdropping than others, and the emissions can typically be captured at ranges between approximately 182 and 304 m (600 and 1000 ft, respectively) from the device using the proper equipment and software.

There are several general classes of side channel attacks:

- power monitoring attack: uses the varying power consumption of a device
- electromagnetic attack: uses leaked electromagnetic radiation
- acoustic cryptanalysis: similar to power monitoring, but uses sound produced by devices during computation
- cold boot attack: also called a data remanence attack, uses a memory dump in conjunction with a hard boot of a computer
- timing attack: uses the variations in timing while performing computing activities
- optical attack: monitors visual indicators like a hard disk activity indicator

Countermeasures include eliminating or reducing the emissions by shielding or disabling external indicators. Another method is to eliminate the relationship between the data and the source of emissions, making the data unrelated to the leaked emissions. Shielding, such as a Faraday cage, is useful to contain electromagnetic radiation, while obscuring visible indicators and providing enclosures that dampen or eliminate sound will serve to protect against optical and acoustic attacks. Algorithms that assemble or translate uncorrelated data from physical phenomena can also be used to thwart several of the more sophisticated attacks described above by providing an understanding of how data are collected and extracted.

3.0 SOCIAL ENGINEERING

3.1 Definition

Social engineering is essentially the compromise of the human asset by various means available to an attacker; it is generally agreed that social engineering is "human hacking." Using a variety of information sources, specifically social media, an attacker can "mine" data from the various sites frequented by a target and build an "attack profile" that is designed around the target individual. Given that many individuals persist in sharing deeply personal or revealing information on social media for the world to see, this method has been and will continue to be a primary means of obtaining information on a target.

3.2 Intelligence Gathering

There are several sources of information on an individual or organization that are useful to a threat actor; these are intelligence-gathering disciplines and are defined by the intelligence community as follows:

- HUMINT: human intelligence
- OSINT: open-source intelligence
- GEOINT: geospatial intelligence
- TECHINT: technical intelligence
- SIGINT: signals intelligence
- MASINT: measurement and signature intelligence
- CYBINT: cyber intelligence
- FININT: financial intelligence

Of these eight disciplines, the first two, HUMINT and OSINT, are typically the most often used to compromise a target. Both rely heavily on the propensity of humans to be helpful or to want to share information for various personal reasons. This characteristic is useful to those engaged in intelligence-gathering activities and is widely used by investigators to determine perpetrators of illegal activities and other investigative tasks such as root cause analysis.

HUMINT is the technique of gathering intelligence by means of interpersonal contact. This differs from the other means of intelligence gathering in that it is typically not technical in nature. HUMINT may be as simple as two people discussing how a process works or discussions with a vendor at a trade show—either, or both, are motivated to obtain information from the other to further their goals; these goals may be perfectly innocent, as in determining if a facility needs an upgrade, or may be nefarious, as in determining what type of firewall a facility is using and how it is configured. In the worst case, the target may actually give up passwords or other credentials because they trust the other party. At the bottom of it, HUMINT is based upon basic interactions between people on a personal level—the threat actor gains the target's trust and then exploits it.

HUMINT can be thought of as a refined confidence game or con. As touched on earlier, these are a distinctive form of fraudulent behavior that are designed to trick an individual into an exchange, whether it be of goods or intelligence, that is not mutually beneficial; in fact, it is rarely beneficial to the target (or "mark" in the parlance). These scams take the form of the "long" and "short" cons—in the former, the operator develops a relationship with the mark over a period of time, which could be weeks or months; in the latter, the operator typically relies on a mark's good nature or naivete to affect a quick response that can be exploited. These techniques are easily carried over into the cybersphere by skilled practitioners who have targeted a specific individual or organization. These techniques work so well because they exploit typical human characteristics of need and greed, dishonesty, vanity, lust, compassion, incompetence, and naivete; this is even before cognitive biases are considered. These characteristics are difficult to train out of people and on a strictly personal level are effective vectors for a determined attacker.

OSINT is a multifaceted approach to intelligence gathering about an intended target. OSINT uses open-source information to determine an attack method and is widely used by the Russian and Chinese intelligence services to develop attack profiles on individuals and organizations. Used in concert with HUMINT and several other techniques listed above, a thorough and detailed attack profile can be developed in a relatively short period. Data obtained from various open sources varies in quality, however, and a competent threat actor will use several sources of OSINT to back-check and confirm information before using it to perpetrate an attack. A primary source of OSINT is social media. The phenomenon of social media has fundamentally changed the way humans interact on all levels; the rise of social media is an inflection point in the development of the species and is not necessarily a positive development as has been seen. A discussion of social media and its relationship to the rise of cybercrime will be discussed further at the end of this section.

OSINT comes in many forms: the aforementioned social media is a primary vector; blogs, general consumption media, professional and academic publications, technical reports, and commercial data are all open-source information that can be found either on the internet or in a public library. Much of this information, taken by itself, is innocent and poses little risk to those who published or developed it; taken as a complete body of research, however, it provides a detailed picture of an individual or organization that can be used to develop an attack profile. A simple example is a typical social media profile: personal information is widely posted to these profiles and this provides a starting point for a threat actor to "follow the trail" to a potential vulnerability. At the very least, it gives the attacker the data needed to determine access credentials or other means to obtain them.

Intelligence gathering is an essential first step in developing the attack profile that will be perpetrated on the selected target. It is essential that any cybersecurity policy address this potential vulnerability and treat it as such. This requires an organization to educate and possibly monitor those individuals who are engaged in developing or operating critical facilities. It is important to remember that any means of communication is subject to interception, whether it is interpersonal or electronic—there is no reasonable expectation of privacy in a public setting—good policy, based upon thorough understanding of the threats involved, will guide administrators in developing effective training and resulting policies.

3.3 Managing the Human Asset

The human asset is the most frequent and easiest asset to compromise. Most successful breaches occur because of incompetence, irresponsibility, dissatisfaction, or some nefarious purpose. Human error is a large factor in many breaches that could have been otherwise prevented. Managing the human asset requires specific skills and what is called emotional intelligence to understand what drives the asset and how to effectively control the asset to avoid an incident. Intelligent cybersecurity policy and enforcement, education, and promotion of ownership and personal responsibility are key to maintaining a secure facility; however, social factors, such as cognitive biases and level of education, along with structural factors, such as labor union rules, can be an impediment to instituting proper cyberhygiene. These can be overcome in time, but it is essential to begin the process, at least on an administrative level.

3.3.1 Cognitive Biases

Cognitive biases are systematic errors in thinking or judgment that occur when people create their own "subjective reality" from their interpretation of the information presented to them. Cognitive biases are not necessarily a bad thing; many cognitive biases serve as a shortcut for decision making in a given situation or context and can lead to more effective actions. Many cognitive biases, however, allow a compromised individual to rationalize aberrant behavior, resulting in harm to themselves, others, or their employer. There is an evolving list of cognitive biases identified from research done over the last 60 years, some of which are described below for the sake of clarification.

Confirmation bias is the most familiar type of bias to most people. Confirmation bias is classified as a belief-based bias and is the tendency of an individual to gravitate toward sources of information that confirm their beliefs, thereby confirming and strengthening those beliefs. The "bandwagon effect" is classified as a social bias and is the tendency of an individual to do or believe information because many other people do or believe the same. Fundamental attribution error is another social bias and is the tendency of an individual to overemphasize a behavior in another person while underemphasizing the same behavior in themselves. Hindsight bias is another belief bias and is the tendency of an individual to see past events as being predictable ("I knew it all along").

The most intriguing cognitive bias in the context of cybersecurity is what is called the Dunning-Kruger effect. The Dunning-Kruger effect (n.d.) is a belief bias and postulates that the incompetent do not have the social skills to know they are incompetent—they tend to overestimate their own ability—while experts tend to underestimate their ability. This has a bearing on cybersecurity and promotion of cyberhygiene. Many individuals in sensitive positions may harbor this bias—they are sure of themselves and bristle at any criticism of their work. Conversely, they are susceptible to flattery and confirmation bias. At the other end of the scale, experts in a field may not be sure of themselves and then question their ability or their judgment; they are also susceptible to flattery, and many do not take criticism well. Both can be compromised by skilled threat actors that play to these biases.

Cognitive biases, in concert with social media environments, can provide a fertile landscape for compromise of a human asset. Threat actors, particularly nation-states that merely observe other cultures to determine these exploitable vulnerabilities, will continue to see these individuals as potential assets.

3.3.2 Promoting Ownership

This is a broad topic, and the focus will be on a few key points to allow further study. Many breaches are perpetrated through what is popularly called a "disgruntled employee." Although this may be true in some cases, disinterested, frustrated, or irresponsible individuals ("disgruntled") can, and do, cause breaches to happen. This type of person will not take ownership or responsibility naturally; they exist in every organization and represent a potential vector for a system breach. This type of employee should never be allowed to access credentials to critical systems because of their inherent resentment, lack of interest, or the overt hostility many of these people feel for their organization or employer. It has been said that there are some people who cannot be reached; there are still many more individuals who will step up and take responsibility if given the opportunity, and, more importantly, clear direction and a reason to perform their duties effectively. Identifying these responsible people and isolating those who present a threat are functions of effective leadership in any organization. Those with a sense of purpose and mission and those who understand the importance of their duties and responsibilities are most amenable to assuming a sense of ownership in the facility.

Some points to consider are as follows:

- Communicate. An effective leader will communicate early and often, particularly about the organization's mission and goals. This helps an individual to understand their purpose in the organization and that they are part of something bigger than themselves.
- Delegate authority. Give key individuals a leadership role, and make it known they are trusted and can make decisions on how to accomplish their tasks. Encourage them to solve their own problems, but hold them accountable.
- Emphasize their role as a thought leader and share critical information about the organization and identified threats.

The key takeaway here is that an individual will take responsibility for the organization's well-being and this will evolve into a sense of ownership—which leads to the individual taking the initiative to protect the organization. Although simplistic in description, this technique is often hampered by a lack of interested or competent individuals. However, even one individual can make a difference in how an organization is protected from, or responds to, a breach.

3.4 Role-Based Access Control

Role-based access control, or RBAC, is a technique used to restrict access to networks or network segments that are used in specialized areas of the organization. For instance: A shift supervisor needs access to the facility SCADA network to observe and adjust the process; they would not need access to the accounting department network. Conversely, the facility's accounting staff does not need access to the SCADA system to do their jobs. Each role in the organization determines the level of access to a network; this approach restricts system access to authorized users. All systems have the capability to restrict access and provide the credentials that will allow access to specifically designated areas or networks.

Role-based access control is effective in limiting the access an individual has and thereby limiting the extent of a breach should those credentials be compromised. Many breaches occur from human error on the office side of the facility; restricting access to the process side of the facility limits or eliminates any compromise that would allow entry into the process control network. Similarly, if a facility worker is compromised, the accounting network is off limits and can be contained to the process side.

Role-based access control is another way to describe hierarchical access control. All SCADA systems have defined levels of access: At the lowest levels, an individual has only rights to observe the process, while privileges accumulate as the level of credential is promoted. To further the example, an administrator-level credential would allow complete access to not only observe, but to modify the SCADA system parameters—only trusted individuals have this level of access as it can be used to completely compromise, if not control, the facility if these credentials are obtained.

3.4.1 Principle of Least Privilege

Another method of restricting access based on role is called the principle of least privilege. Whereas RBAC dealt with levels of access to specific networks or network segments, "least privilege" is a method to restrict access to only the information and resources the individual needs to perform their job or intended function.

3.4.2 Background Checks

With the rise of cybercrime and cyberwarfare, it is not unreasonable to presume that a determined threat actor could plant individuals within an organization as a "mole" to be activated at a strategic time. Also, an individual can be compromised because of issues such as illegal or immoral behavior, gambling or other debt, or inherent criminal tendencies. Given that thorough background checks that include criminal records are easily and inexpensively available, this is a simple screening technique that will serve as a first line of defense when hiring.

3.5 Social Media and Cybercrime

Cybercrime is a low-risk and high-payoff activity. As has been seen over the last several years, the incidence of cybercrime is on the rise. Most of these crimes are perpetrated by the compromise of the human asset. As has been shown, much of the information used to design an attack on an individual is gleaned from OSINT, of which a major component is social media. Social media is a rich source of information about individuals and organizations that can be thoroughly sifted and collected for nefarious purpose. People are often and easily compromised, most of them unwittingly. Much has been written about the mechanisms of social media; it must be understood that the primary product of social media is your personal data. The trick is to provide a venue that promotes the posting of this information to allow harvesting and sale to interested parties. This model is the perfect scenario for threat actors who use this information, not for targeted sales or political campaigns, but for developing an attack profile that can be used to compromise an individual and/or their employer.

One particularly insidious technique is the "like." To most people, this is an innocent game that substitutes for human interaction, the venue providing quasi anonymity that in turn promotes more interaction, whether good or bad. Getting liked, particularly by those with cognitive biases or lacking in social skills, provides an endorphin rush similar to other sources of pleasure—as with chocolate, being liked on social media is addictive. The addiction drives the person being liked to take the exchange further; like all addictions, it changes brain function and behavior, bypassing normal precautions someone would ordinarily take when conversing with a stranger. The lack of physical cues that would excite caution in a personal exchange are not present in a virtual exchange, and virtual anonymity removes many inhibitions that would prevent self-destructive behavior in the real world. Those lacking in social skills, the socially isolated, the "loner"—these are some of the most susceptible to this technique of manipulation. A skilled operator can play this addiction for gain; those susceptible to flattery and compromised by endorphins will readily give up personal or privileged information when "in the moment." This discussion is far beyond the scope of this Manual of Practice, but it should be noted that this sort of behavior is hazardous and those who are addicted to social media pose a potential risk to a critical facility. Further reading on this complex subject is listed in the Suggested Readings section at the end of this chapter. It is worthwhile to explore ways of educating staff to what constitutes sensitive information and why it should not be posted online. There can be a balance without resorting to monitoring an employee's online behavior; however, the pull of social engagement online is powerful and can overwhelm common sense and responsibility.

4.0 PREVENTION TECHNIQUES

Cyberhygiene is a discipline that must be taught and enforced; it is the first line of defense against cyberattack. As has been discussed, the human element is the most easily and most often compromised. Social engineering attacks are so prevalent and so popular because they work and provide good returns. To reiterate, the first line of defense against a cyberattack is through the proper training of staff and the setting and enforcement of policies designed to take guesswork and irresponsibility out of the equation.

The following section will discuss technical countermeasures that should serve as a guideline for the structuring of a robust cybersecurity program. The chapter concludes with a discussion of policies that can be used to enforce good cyberhygiene, particularly password generation and usage.

4.1 Defense in Depth

4.1.1 Definition

Defense in depth is a term borrowed from the military that describes methods of layered security and defense. The model could be thought of as an onion; as you peel away one layer of defense, another is presented. Layered security, or defense in depth, constitutes a continuous defense system, each succeeding layer presenting a more robust defense method than the last.

4.1.2 Implementation

Implementation of the model varies greatly depending upon an organization's budget and the value of the assets that can be attacked. The goal of the model is to frustrate, confound, misdirect, capture, and ultimately defeat the attacker. The use of network intrusion detection and prevention systems (IDS/IPS) is integral to a robust network defense. A typical layered defense model will use all or some of the layers listed below (from the public-facing attack surface inward):

- Perimeter defenses: These can be physical barriers; doors, fences, guards, patrols, dogs. The goal is to physically block the attacker from accessing a facility. Firewalls are a primary method of network perimeter defense. A firewall actively examines and filters incoming data to allow or disallow passage to the destination within the protected network based on source IP address or access credentials. Firewalls will be discussed in detail below.
- Network defenses: Including the firewall mentioned in perimeter defenses, which only work at the point of entry into the network, network IDS/IPS are often the first line of defense after the failure of a human asset. These systems monitor traffic within secure networks and identify possible intruders; advanced systems will contain and eliminate a threat. Network traffic is analyzed for attack signatures,

and patterns of data traffic are "learned" by algorithms used by the detection systems to determine what is legitimate network traffic from recognized users. Employees who suddenly step outside of their normal behavior or access networks that are not ordinarily part of their job function can be flagged and segregated. System administrators are alerted, and the activity is assessed to determine if there is a threat. Logging of data activity from a suspected breach or suspect employee is also a useful tool to prevent or contain a breach. Monitoring of mobile devices is a particularly useful function of activity monitoring systems, and wireless intrusion detection is a mature technology that actively scans for intruders and has the sophistication to misdirect or eliminate the intruder.

- Host system defenses: Also called "endpoint" protection, this type of protection is either custom-built or consumer-grade "virus" protection software. This software can be stand-alone, as in operating on one computer connected to the network, or it can be a robust network solution that is integrated with an organization's cybersecurity infrastructure. This type of system is effective and can flag potential problems down to the file level—if an employee downloads an infected file, the system communicates back to the system administrator, who takes the appropriate action, like isolating the computer from the network. This type of system is also useful for preventing employees from visiting blacklisted websites or for logging their activity when they do in order to determine the level of threat.
- Application-level defenses: These defenses operate at the application level of the open systems interconnect (OSI) network model; the application level is where a user interfaces with the system and is at the top of the "stack" that defines and structures network communication systems. Application-level defenses control how users interface with the applications that are used to access system resources. These include passwords and multifactor authentication for primary access. Defenses at this level also include how a user can use the application program, what resources the user can access, and what rights are granted to the user. Hierarchical and role-based access control (RBAC) are used at this level (described below). These techniques allow access only to resources or systems that the user needs to do their job functions and can isolate a breach or limit it if properly configured.
- Data defenses: Breaches happen in even the most well-defended systems for a variety of reasons. One method of data defense is to move the data offline on a regular basis. This means removing the data

from the actual physical premises; these are commonly known as "backups." Backups are traditionally done to preserve critical data to protect against hardware failures; hard drives crash, software becomes corrupted. Fires, floods, and other disasters can wipe out an organization's data within a very short time. Although data are typically safe because of nightly backups, the same cannot be said for server or network operating systems and the configuration parameters programmed into them. Every operating system is customized to some extent by the end user to suit their particular needs—no system works for everyone out of the box. In a well-known example, all an organization's servers were corrupted by ransomware; it was by a stroke of luck that one server in a remote office was offline because of a power failure. This server contained the only uncorrupted, in fact, pristine copy of the server operating system; it was what saved months of reconfiguration, bringing the company's network back up in weeks. In addition to backing up data and moving them off-site, pristine copies of key operating systems must be copied and moved to secure locations upon installation and after any customization. If a system is compromised and held for ransom, the system can be wiped clean and brought back up quickly.

4.2 Security Controls

Security controls reduce your attack surface and will frustrate an attacker. As discussed above, defense in depth requires the use of several forms of security controls to be an effective barrier to a cyberattack. Security controls fall into three broad categories: physical, logical, and administrative controls. Each is unique to the others and provides specialized protective functions based on facility type and the value and nature of the asset being protected.

4.2.1 Physical Controls

Physical controls are exactly that—a physical barrier that prevents access to an asset. This type of protection is typically used as a perimeter defense system. Fences and walls are the most common and obvious method of preventing access to a facility. It is not uncommon for high-value facilities to have double fences and razor wire barriers at the physical perimeter of the facility. An enhancement to this system is infrared or radio frequency intruder detection systems either between the fences or adjacent to the fence in the warning track or "kill zone," either inside or outside the protective perimeter. Guards and patrols are also used by high-value facilities for either continuous monitoring or rapid response to a breach. All facilities of this type would use manned access points to screen personnel, visitors, or contractors. This is the high end of the scale and can be costly; however, this type of protection should be considered in proportion to the risk of loss or compromise by an intruder. At the lower end of the scale, fences, walls, and automated entry systems based on access credentials presented through a keypad or by an access card are typically sufficient for a typical facility. Again, physical security methods and the scale of these protections should be instituted on a scale that is proportional to the value of the assets being protected.

Wireless and closed-circuit video surveillance is a relatively inexpensive but effective means of monitoring a facility perimeter, as well as allowing surveillance of sensitive or key areas of a facility. Wireless technology is mature but can be jammed by a determined attacker; wired cameras are more secure depending upon method of installation, but can be rendered inoperable by lasers, spray paint, and even a hammer. Installation of these devices is just as important as their placement; overt placement of the camera is a warning, while covert placement is protection. Placing a surveillance camera in an inaccessible or hidden location or disguising it as something else is a proven method of deceiving an adversary. As in all things cyber, if the attacker cannot find the asset, they cannot attack the asset. Video surveillance is widely used for process monitoring in dangerous or unpleasant process areas, and complete systems are well within most facility budgets.

Finally, physical controls can be as simple as locked doors and cabinets. A sensitive area, such as a server room or telecommunications closet, can easily be significantly damaged and disabled by an attacker simply by entering through an unlocked door; of course, this is assuming the attacker has gained physical access through an unsecured perimeter, which is the first line of defense. Leaving critical areas vulnerable to attack because of a failure to provide a locked cabinet or doorway is poor security practice and can lead to critical systems being disabled for weeks.

4.2.2 Logical Controls

Logical controls consist of devices or software that provide defense at the data level. These defenses come in many different forms; among the most common logical controls are firewalls, network segmentation, misdirection, proxies, and DMZs (demilitarized zones). Briefly, logical controls actively examine data either entering or leaving a network and act based upon rules (as in a firewall) or prevent the mingling of secure and less secure data (as in segmentation). The goal of logical controls is to defend a secure network against logical attacks that use network vulnerabilities to accomplish a breach. Another goal is to reduce a facility's attack surface, making its network presence as insignificant as possible—an example would be the stealth bomber; its technology reduces its radar signature to that of a hummingbird.

4.2.3 Administrative Controls

Administrative controls are organizational defenses that are effected using policies, enforcement, training, and the institution of strict access hierarchies and separation of access privileges. In the first case, the most obvious and vital policy that should be instituted is a robust password policy; password policies and their enforcement will be described below. Training is an important component of cybersecurity, and the training of personnel in proper cyberhygiene will allow the development of a sense of ownership in the facility by key personnel. Often staff does not have "buy-in" to cybersecurity policy and see it as a chore; this leads to carelessness and even willful disregard of security policies. Fostering ownership in a facility or even in a process area gives the employee a sense of purpose and responsibility that translates into ownership of that task or facility, which the employee wants to protect. Training is an important complement to physical and logical security methods.

Hierarchical access control simply means that access to sensitive or secure networks, or to privileged information is based upon a ranking of privileges that increases in the breadth and scope of access as credentials escalate. On the lowest levels, for instance, an employee has access rights that allow them to log into the system but only view a process; the employee cannot change or configure anything. Moving up one level, an employee could be granted privileges to change process control parameters or set points, pump speeds, or point levels for tank level control. This typically requires the employee to have a higher level of training and experience. This employee cannot change any algorithms or graphics—this could be considered a supervisor-level access credential. Moving up another level, an employee could be given full access rights to every aspect of a system, including updating, patching, and processing algorithm control parameters; this is an administrator-level access credential. It should be remembered that an attacker will try to compromise an employee with any credential level but will always try to escalate their access credentials once inside a compromised system.

4.3 Zero Trust Architecture

Zero trust is rooted in the principle of "never trust, always verify" and is used in concert with network segmentation and obfuscation to effect security at the user level, specifically at OSI application layer (Layer 7) where the user interacts with the system. The underlying concept is that trust is a vulnerability. Zero trust architectures are used to achieve granular visibility for all traffic—to and from users, network connected devices, local and remote locations, and applications. The zero trust model considers the entire network to be potentially compromised and hostile. In fact, the model dictates that even though actors, systems, or services have satisfied the protocols for perimeter access, they should not automatically be trusted. This is based on the premise that perimeter defenses are not working. It goes further by stating that once a threat actor has breached the perimeter, they are free to move horizontally and vertically through the system and escalate their privileges as they go. The model calls for microsegmentation of network resources within the network and particularly for sensitive systems.

Zero trust models rely on techniques like multifactor authentication and push notifications (sending a verification code to a mobile device) to provide authentication. Multifactor authentication is described later in this section.

4.4 Reducing the Attack Surface

4.4.1 What Is the Attack Surface?

Simply put, an attack surface is the entire public- or external-facing entry point into an organization's systems or networks. Attack surfaces include digital, physical, and social engineering attack surfaces.

Some examples of the attack surface would be the organization's website, open ports on an external-facing system, employee-only portals, and unsecure virtual private network (VPNs) or remote access software. Combined with stolen or illegally obtained credentials or other access information, an attacker can systematically probe these entry points into the network and exploit a vulnerability. Aside from developing and cultivating a culture of cyberhygiene, reducing the attack surface is key to a robust defense posture.

4.4.2 Reducing Online Presence

The first step in reducing an organization's attack surface is to reduce its online presence. This may be counterintuitive to many commercial organizations that want to maximize their online presence so that they can be found easily. This drives traffic, which theoretically drives sales. However, a critical facility like a water treatment facility has no need for a large and splashy website to drive sales. There are specific ways to reduce your online presence and, in so doing, reduce your external-facing attack surface. It is important to remember that these methods must be integrated within a culture of cyberhygiene that includes good password policies and the several technical methods of protection and cyber defense described in this chapter. It is also important to realize that there is also an internal attack surface that must be considered—employees can also be a threat—and this possibility must be considered.

There is no need for a utility website to have any connection or access to the process control network, and the website should be a stand-alone and informational offering. The website, if absolutely necessary for public information purposes, should be hosted separately and if possible, with a different domain name. This will also misdirect an attacker to the website, while the entry portal, through a firewall for authorized users or vendors, is obfuscated. If an organization uses a website that is integrated into the overall enterprise network, then steps need to be taken to segregate the external-facing portals and associated web applications from the secure assets; this goes for any secure asset and not just a process control network. Some basic steps are to minimize the user experience and not give the user or attacker many options for interaction with the website; do not optimize the website for mobile use; do not provide valuable information or information that can be easily discovered by search engines. By no means get listed on web directories that expand your visibility and make it easier for search engines to find your organization.

Periodically, search for your facility and see what comes up. Examine the search results for possible leaks of information or any means of possible access into the secure network. This is an easy test to perform and should be done regularly. It is also good practice to do a periodic search on key employees to determine if any access information or other detailed information is being shared that could compromise the employee or help an attacker determine access credentials or where a hidden network portal may be located. In most cases, Google can be used to uncover what information about your organization is public and what an attacker can find out without the need for advanced tools, social engineering, or reconnaissance. These searches will allow you to patch the holes in your external-facing attack surface before they become a useful exploit.

Finally, avoid social media and delete any accounts that may possibly be used as a pathway into the secure network. As was discussed in Section 3 of this chapter, social media is the primary means by which reconnaissance is done to develop attack profiles. Delete or deactivate any old accounts that may still be active from former staff; this is often overlooked, and many accounts contain passwords that can still be used for exploits. Personal data can be hidden for key staff by a variety of means, including using aliases. This is extreme, however, and can lead to further confusion and access problems for legitimate employees.

4.4.3 Methods of Reducing the Attack Surface

Other common methods of reducing your online and data attack surface will be described below; this list is not exhaustive but will serve to direct the reader to further research.

4.4.3.1 Deception, Misdirection, and Honeypots

Misdirection and deception are perhaps the oldest methods of protecting valuable assets. Using deceptive information to direct an attacker to a dead end, or to cause the attacker to expend valuable resources trying to find an exploitable target, will eventually cause them to move on to a softer target. A common technique is to set up what is called a "honeypot," which is a decoy and sacrificial computer system that an attacker is misdirected to by deceptive information. It mimics a valuable target and causes the attacker to expend resources looking for vulnerabilities. This method directs the attacker away from other targets, which can be further protected by obscuring them. Honeypots are used as decoys to allow network administrators to monitor and track attackers and gain valuable information about them without exposing the target network to compromise.

4.4.3.2 Network Obfuscation

Obfuscation means to obscure something; in other words, obfuscation changes the outward appearance of something to deceive an attacker. Obfuscation is used on both sides of the cybersecurity equation. Websites, ports, or other externally facing entry points can be obscured by changing their digital signature to make them look like something else. Obfuscation is another old technique for protecting valuable assets. Obfuscation is also used by malware to make it appear to be something else; PDF obfuscation is a common technique for the delivery of malware. The use of obfuscation is a broad topic that deals mostly with the computer code that makes up web applications or other common programs.

Network obfuscation techniques also use techniques of virtual obfuscation, which works by routing data through a series of network nodes called "onion routers" and encapsulating data in a series of encryption layers (hence the name). These routers know only the previous and next routers for a given data packet, decryption happening when the message arrives at the destination node. This method effectively prevents an attacker from tracking data through a network and effectively obscures the target destination.

4.4.3.3 Demilitarized Zone

A demilitarized zone (DMZ) is a perimeter network that protects an organization's internal private network from untrusted or unauthorized traffic. A DMZ sits between an external- or public-facing network entry point and the private local area network (LAN). If an organization must have a public website to serve its business purpose, then a DMZ is essential. A DMZ can be used to provide services to the public without allowing access to internal networks and can include such services as web and mail servers. It also prevents reconnaissance by preventing access to the internal LAN. Ideally, the DMZ is located between two firewalls, but a single firewall can be used effectively, and is the most common configuration. In a single firewall setup, one interface is to the public internet, another connects to the internal LAN, while another connects to the DMZ. Deploying a DMZ between two firewalls is a much more secure arrangement, as would be expected. The first firewall allows traffic to the DMZ only; the second firewall allows traffic from the DMZ into the internal LAN. Both arrangements use rules, access control lists (ACLs), and white- and blacklists to monitor and control traffic and determine which traffic is allowed access to the internal LAN.

4.4.3.4 Closing Open Ports

Open ports on a server or other externally facing equipment is an invitation to an attacker. A port is a virtual communication channel that is used to provide a particular service. Ports are software based and are typically managed by the computer or server operating system. Data communication protocols are associated with specific ports and are configured to accept data packets transmitted with particular protocol. There are 65 535 ports available for communication between processes and to provide specific services. There are "well-known" ports, numbered from 0 to 1023, that provide the most common computer services; Port 80, for instance, is the most common well-known port and identifies web traffic for a web server (Port 443 is the secure version). Ports 1024–49151 are registered ports and are used for the convenience of the internet community and are not assigned to a specific service or controlled. Port numbers from 49152 to 65535 are called "dynamic ports" and are not assigned, controlled, or registered. Ports 1024 through 65535 are also called "ephemeral" ports because they are typically used only for a specific communication session. Some are used by equipment manufacturers for exclusive access to their equipment.

A closed port is a safe port. If there is no need to have a port open, then best practice is to close it. A common technique to discover a system's vulnerability is to perform a "port scan" to find open ports. This technique is analogous to walking down the street and turning doorknobs to see which door is unlocked. Running a port scan will reveal which ports are listening (receiving data) and what security devices reside at those ports (a process known as "fingerprinting"). Port scanning is a favorite tool of attackers but is also used by penetration testers to test systems for vulnerabilities. A free program called Nmap can be run to determine port status on any connected system.

4.5 Penetration Testing

Penetration testing, or pen testing, is a method of ethical hacking. Pen testing attempts to penetrate an organization's network defenses as would a malicious attacker to determine vulnerabilities in those defenses. It is essentially a simulated cyberattack perpetrated by "white hat" hackers who specialize in breaching network cybersecurity at the behest of the owning organization. Among the goals of the pen test is the identification of open and unused ports, tweaking and refining firewall access rules, and generally eliminating security holes throughout the system. Websites are also subjected to simulated attacks to determine common web application vulnerabilities. Finally, pen testing is used to gain access to secure systems and then escalate those privileges as deep as the pen tester can manage. Pen testing is designed to discover and mitigate any security issues that a skilled attacker would be able to find and exploit.

Common techniques involve both external attacks on the network perimeter and attacks from within the network itself to determine where an attacker could go if the perimeter was breached. In some cases, the pen tester goes into the test "blind" without any information on the target network, simulating the actions of a real hacker; this can be done with or without the knowledge of the employees, with only top management being aware of the simulated attack. Pen testing can be manual or automated; automated pen testing can be continuous while manual testing involves an intentional simulated attack. In all cases, the goal is to gather information about system vulnerabilities and report back to the organization's IT security team.

4.6 Red/Blue Team Exercises

A red team versus blue team exercise is another method of determining an organization's security vulnerabilities and involves simulated attacks to accomplish this. The red team are the attackers and use ethical hacking techniques to attempt to breach the network defenses. The blue team are the organization's incident response team that responds to and defends from the simulated attack. Conducting these exercises allows an organization to test its network defenses actively and rigorously in a low-risk environment, and with the full knowledge of the system administrators and organization management.

After the exercise, the red and blue teams will discuss the methods of attack, the countermeasures used in defense, and the responses that were observed. There are often time limits or targets for the blue team to discover and contain or eliminate a simulated attack. The red team will use a variety of methods, and if the blue team experiences difficulty in detecting and containing the attack, the effectiveness of the attack response is analyzed, and new countermeasures or mitigation are developed. The red and blue teams typically collaborate on the attacks and response methodologies; this collaborative approach is called a purple team exercise.

4.7 Network Segmentation

4.7.1 Office Versus Process

In most facilities, a process control network is in place to monitor a process monitoring and control network, while another is in place for use of office staff for business purposes. In some cases, particularly in facilities with older equipment or limited budgets, these networks are combined and typically share an internet connection. In all cases, best practice is to segregate the office network from the process network. There are several reasons for this, most of them being related to cybersecurity, while others pertain to network performance; this discussion pertains to cybersecurity.

In essence, combining an office and process control network doubles a facility's chances for a breach. The reasons for this are easy to see: An attacker can be from within or without the network. Many breaches are perpetrated by unskilled or disgruntled employees. Those not trained in process network operation can do significant damage if allowed to access any part of the network—this could inadvertent or intentional. It should be understood that an office network that is integrated with a process network could harbor any manner of malware and infect a process network; the infection could then be passed back and forth between them even though IT has done its best to reconfigure the compromised network.

Important and high-value network resources, such as critical process control and SCADA servers, should never share network resources with an office network. Another definition is that IT, which is traditionally business oriented, and operational technology, which deals primarily with industrial and machine communications, including SCADA, should not coexist on the same network segments or share network resources, such as servers, firewalls, and internet connections. This is a basic design principle when contemplating a process control network. For organizations with limited budgets, it is a false economy that can compromise an entire organization's ability to function, let alone control a critical process. Each network segment should stand on its own; each should have a well-configured firewall and separate internet connection. There should be no cross connection between the two and data sharing should be thoroughly controlled through removable media, which should be monitored and controlled. Segregating these networks is a basic method of network defense and conforms to the principles of access control discussed above.

4.7.2 Firewalls

A firewall is a barrier between a secure network segment and an unsecure network segment. The purpose of a firewall is to limit or eliminate traffic between an unsecure network segment and the secure network segment. Traffic from the secure side is similarly controlled; this traffic may be the result of a compromised computer within the secure network that is attempting to establish a communications channel to an attacker's system outside of the organization. A firewall is, aside from complete air-gapping, a primary means of network perimeter defense. It can be thought of as a physical barrier to unwanted or unauthorized entry. Typically, firewalls are used to separate the public-facing side of the network from the secure side, such as standing between the internet and a process control network. Firewalls can be used at any level of the networks to isolate network segments from one another within a larger organizational network. Firewalls can be hardware or software based, or both. Many consumer-grade wireless routers employ software-based firewalls to filter incoming traffic. Commercial firewalls are almost always hardware devices that provide robust protection and may also have virtual private network (VPN) capabilities for added communication security. Cloud-based firewalls are available as a firewall as a service (FaaS), like software as a service (SaaS); these firewalls work the same as other firewalls except they do their work in the cloud.

The two primary methods used by firewalls to determine access are internet protocol (IP) filtering and access control lists (ACLs). Access control lists are simply listings of permissions used by any of the above types of firewalls to specify which users or system processes are allowed access to specific computers, networks, or systems. Firewalls operate using specific and user-configurable ACL rules. A rule will typically include the source IP address, the communications protocol being used, the port number access being requested, and the destination IP address. Only trusted sources will be allowed access to the secure network (IP source address to destination address), or only that traffic using a specific protocol. Once the source is allowed into the secure network, the firewall ensures that the traffic is allowed to access only a specific port, depending upon the source's IP address. A privileged user can be allowed access to any destination and port, while restricted users can only access specific destinations and ports. This is essentially how a firewall works.

Firewalls are classified by their functionality: Packet-filtering, proxy, network address translation (NAT), and web application are some examples of specialized firewalls. Packet-filtering firewalls are the most basic type, examining incoming packets and allowing or disallowing their passage based on firewall rules. Proxy firewalls operate at the application level by performing stateful and deep packet inspection. NAT firewalls operate by assigning a public IP address to a group of private IP addresses within a private network, essentially hiding those addresses from scanning for and attacking them. Web application firewalls are used for monitoring, filtering, and blocking data as they flow into and out of a website or web application. There are several other types of specialized firewalls that are beyond the scope of this manual; the reader is encouraged to pursue further research into this important topic.

4.7.3 Virtual Private Networks

Virtual private networks (VPNs) are a means of secure communication between two computers, networks, or systems. Virtual private networks extend a private network across a public network and act as though the remote computer is connected to the local network. Unlike firewalls, VPNs do not filter traffic for malicious content; if the malicious content is present in the transmitted data, it too will be transmitted within the secure communications channel. A VPN can be thought of as a secure communications channel within a publicly visible communications channel—a tunnel within a tunnel, with the inner tunnel encrypted to prevent interception. A VPN is an enterprise network that uses the public telecommunications infrastructure, just as the internet does, but establishes private and secure connections over an unsecure network. A private communication network uses private leased lines in this context, whereas a VPN uses the public internet.

Data transmitted through the VPN are encrypted at the network gateway before traveling through the public internet. The data are encapsulated within a data packet and sent to the destination gateway. The gateways protect the networks at either end from intrusion from the internet. The data packet is encrypted end-to-end but has an open header that contains routing information. This is called tunneling; once the VPN is set up and connects, the tunnel is set up and the data are transmitted. Tunneling is the technique of encapsulating the data frame in a new packet, which allows the data to tunnel through the intermediate network; a tunnel is a logical path through which the packets travel through the public network.

The common types of VPN are the remote access VPN and site-to-site VPN. The remote access VPN connects a user to a remote private network, allowing access to all resources and services at the remote network. This type is typically used by an employee needing to access the company servers or networks from the road. A site-to-site type of VPN is used to connect multiple remote office or facilities together into one virtual network. The connection is seamless, and the user may access any resource in any office as if it were local. This type of VPN can be of the intranet or extranet type: The former creates a virtual network within an organization, whereas the latter is used to connect one organization to another.

4.8 Virtualization

In the simplest sense, virtual machines and virtual local area networks (VLANs) are software abstractions or constructs. They provide a virtual environment for running tasks or effecting communication by creating a virtual platform or network resource. These virtual devices can support individual computer processes or communication channels or complete server implementations depending upon the level of software abstraction. In essence, the virtual server is a computer program running on a specialized device that allows physical interface to real-world networks and computer peripherals. The difference between a real machine or network and a virtual environment is the ability to completely configure the virtual machine or network, as one would an application program. This ability makes virtual network resources extremely flexible and secure. Isolation between computing or communication processes is enforced because the software construct does not allow direct access to resources that the supporting physical system provides. Access to the physical environment by the virtual environment is controlled and enforced as to what a process, user, or communication channel can access.

Multiple virtual machines can exist in one device, each virtual machine providing a full replica of its physical counterpart; each virtual machine can run its own operating system. In the virtual machine environment, the virtual machine monitor (hypervisor) runs at the highest privilege level and controls the available resources the virtual machines can access. Should a virtual machine be compromised, it can be shut down and a new iteration put in place immediately and seamlessly. From a cybersecurity aspect, data pathways in a virtual machine are very difficult for an attacker to determine because of the nature of the virtual access to system resources, which, unlike physical devices, are not hardwired to a port or network segment.

Virtual local area network are similarly software abstractions that can be configured to provide isolation between classes of traffic. In physical networks, different classes of network traffic commingle, sharing network resources, and can result in undesired access to sensitive data. Ideally, different classes of traffic would be physically isolated from one another; however, this would require redundant cabling, which may not be feasible because of physical or budgetary limitations. The same sort of isolation can be achieved by using a VLAN. A VLAN separates traffic by setting up individual virtual communication channels for network traffic using a device's media access control address at the OSI model Layer 2 interconnectivity. Virtual local area networks operate at OSI Level 2 and provide segregation between logical devices or workgroups. Virtual local area networks can be made very secure by configuring a switch to allow a given device to talk to only other devices on the same VLAN or to devices on another trusted VLAN. The communication channels are completely controlled and serve to segregate sensitive or secure communications from general traffic on the physical network.

4.9 Password Policy and Enforcement

4.9.1 Establishing Policies

A password policy is a set of rules governing the use and construction of secure passwords. These policies can improve data and facility security by establishing a basic set of rules that staff can follow to help them construct and use secure passwords. The policy also specifies how often the password must be changed, and if and when a previous password can be reused. A password policy can also describe the response to a password being compromised and what happens to the user account and associated access credentials as a result. The policy can specify a password blacklist, which is used to deny the use of common or weak passwords. Other things to consider are prohibiting the use of the same password by multiple users, or for multiple systems or devices. An account lockout should be specified, in that after three unsuccessful login attempts, the login application is locked out for a specific amount of time. Inactive account lockouts should also be considered to avoid leaving unused but otherwise valid and active accounts available for discovery and compromise. Inactivity lockout protocols should include the action taken when there is no activity for a specific amount of time, from requiring a password to log back into the system or computer to deleting the account after several months of inactivity.

4.9.2 Enforcing Policies

A password policy can be advisory or can be mandatory. An advisory policy is based on the honor system and trust. Employees can be advised of proper password practice and educated to the security issues that will ensue if poor passwords are used, and how easily common passwords can be guessed. This is a part of fostering ownership as described elsewhere. Mandatory policies can be enforced most readily through system applications that are configured to monitor and control employee password setup and login activities.

This type of enforcement takes much of the guesswork out of the process and eliminates the need for human interaction and administration. These systems also log all logins and login attempts and may be part of or coordinated with intrusion detection systems.

Enforcing advisory policies becomes tricky and often requires intervention of company human resources to resolve a reoccurring problem with an employee. This also requires some sort of monitoring of logins and password construction by a human operator or by an application—either introduces costs and may lead to punitive measures and even litigation. Using system-based enforcement is impersonal and applied to the entire workforce, as well as any vendor or contractor who may be permitted to access the system. This method is also the most secure and provides usable metrics to allow an organization to track and pinpoint a problem or the source of a breach.

4.9.3 Password Best Practices

A basic tenet of password construction is that longer is stronger. A minimum of eight characters is recommended for any password, and 10 characters is better. Passwords such as "1234" or "password" are guessed or cracked within milliseconds by sophisticated algorithms used by attackers. These passwords may be convenient and easy to remember, but they provide no meaningful security. A password is often the first line of defense and protects not only the user but any other network or system they connect to. The password should not only be long, but also contain mixed-case letters, numbers, and special characters.

Mnemonics, or memory shortcuts, can be used to remember or construct a password. The use of a passphrase is a useful mnemonic; for instance: "M@ry_h@d_@_l1ttle_l@mb" is a secure password and easy to remember. Of course, using the same password or passphrase for everything introduces another security risk in that if the attacker gains one's credential, it is a logical next step to try it on every other machine or system associated with that individual. Another strategy is to use a formula that changes with every login based on some characteristic of that site or account. An example would be some standard header and trailer known to you with the special characteristic wrapped inside—this also poses a risk like the last example but would require much more work on the part of the attacker as the credential would naturally change size and content on each iteration, creating a unique credential for each.

4.9.4 Password Blacklists

Password blacklists are kept by organizations and are a list of disallowed words, phrases, or names that are prohibited from being used as part of a password used to access that organization's network resources. These blacklists are used to prevent a user from constructing a password using anything on the list. A good password policy will mandate the use of a password blacklist as part of an overall cybersecurity policy.

4.10 Multifactor Authentication

Multifactor authentication, referred to as MFA or 2FA, is an access technique that uses multiple items to allow access to a system instead of a single password. Multifactor authentication is composed of something you know, something you have, and something you are. Something a person knows is a password or passphrase. Something a person has is a code sent to a mobile phone or email, or it can be a token with a one-time code that is good for a defined amount of time. Something a person is can be a fingerprint, handprint, or retina scan. Used in combination, this access method is very secure when used properly. A combination of identity factors is unique to a particular user and confirms their identity at that moment.

One common example is when asked to enter a ZIP code when using a credit card at a gasoline pump. Online retailers are turning to MFA more often by sending confirmation codes to a mobile phone when a user attempts to log in.

4.11 Updates and Patches

Updates and patches are essential for a computer or system to remain viable and are closely related to endpoint or host-level security. Updates are periodic revisions of software that allow a system to operate efficiently and take advantage of the latest peripherals and communications protocols. Updates are typically supplied by a software vendor, though modern subscription models sometimes use a tiered system of pricing to allow a consumer to have a menu of options for services provided in the periodic upgrades. Operating systems upgrades are typically provided for the life of the product because without them the operating system would crash, and this would have a negative effect on reputations and sales. Updates include security patches that fix security flaws, such as zero-day vulnerabilities. These patches contain the latest attack signatures for the viruses and worms and other threats that are developed every day. The patches also protect an organization and other users from a breach because the patch stops the attack at the target computer and contains it.

Unpatched computers are very vulnerable to cyberattack. Older operating systems, which may not only be unpatched but also unsupported, should be retired or completely updated with new operating systems. These computers present a glaring vulnerability to the owning organization, and budgets need to be adjusted to allow periodic replacement and premium service in order not to introduce a vulnerability that can easily be avoided.

5.0 REFERENCES

BBC News. (2004, April 20). *Passwords revealed by sweet deal.* http://news.bbc.co.uk/2/hi/technology/3639679.stm

Dunning-Kruger Effect. (n.d.). In *Wikipedia.* https://en.wikipedia.org/wiki/Dunning%E2%80%93Kruger_effect

Exec. Order No. 13,636, 3 C.F.R. (2013). https://www.govinfo.gov/content/pkg/CFR-2014-title3-vol1/pdf/CFR-2014-title3-vol1-eo13636.pdf

6.0 SUGGESTED READINGS

Alter, A. (2017). *Irresistible: The rise of addictive technology and the business of keeping us hooked.* Penguin Press.

Benton Institute. (2016, December 2). *Report on securing and growing the digital economy.* https://www.benton.org/headlines/report-securing-and-growing-digital-economy

Capano, D. E. (2019, November 6). *The human asset in cybersecurity.* Control Engineering. https://www.controleng.com/articles/the-human-asset-in-cybersecurity/

Congressional Research Service. (2020, December 18). *Defense acquisitions: DOD's cybersecurity maturity model certification framework.* https://crsreports.congress.gov/product/pdf/R/R46643/2

Department of Homeland Security. (2020, October). *Homeland threat assessment.* https://www.dhs.gov/sites/default/files/publications/2020_10_06_homeland-threat-assessment.pdf

DiResta, R., Shaffer, K., Ruppel, B., Sullivan, D., Matney, R., Fox, R., Albright, J., & Johnson, B. (2019). *The tactics and tropes of the internet research agency.* New Knowledge. https://digitalcommons.unl.edu/cgi/viewcontent.cgi?article=1003&context=senatedocs

Gilman, E., & Barth, D. (2017). *Zero trust networks: Building secure systems in untrusted networks.* O'Reilly Media.

Greenberg, A. (2019). *Sandworm: A new era of cyberwar and the hunt for the Kremlin's most dangerous hackers.* Doubleday.

Hadnagy, C. (2010). *Social engineering: The art of human hacking.* John Wiley & Sons.

Institute for Security and Technology. (2021). *Combating ransomware: A comprehensive framework for action: Key recommendations from the ransomware task force.* https://securityandtechnology.org/wp-content/uploads/2021/09/IST-Ransomware-Task-Force-Report.pdf

Jamieson, K. H. (2018). *Cyberwar: How Russian hackers and trolls helped elect a president: What we don't, can't, and do know.* Oxford University Press.

Langner, R. (2013, November). *To kill a centrifuge.* The Langner Group. https://www.langner.com/wp-content/uploads/2017/03/to-kill-a-centrifuge.pdf

Macaulay, T., & Singer, B. L. (2011). *Cybersecurity for industrial control systems: SCADA, DCS, PLC, HMI, and SIS.* Auerbach Publications.

Mitnick, K. (2017). *The art of invisibility: The world's most famous hacker teaches you how to be safe in the age of big brother and big data.* Little, Brown and Company.

Mitnick, K. D., & Simon, W. L. (2003). *The art of deception: Controlling the human element of security.* Wiley Publishing, Inc.

National Institute of Standards and Technology. (2018, April 16). *Framework for improving critical infrastructure cybersecurity.* Version 1.1. https://nvlpubs.nist.gov/nistpubs/cswp/nist.cswp.04162018.pdf

Nieles, M., Dempsey, K., & Pillitteri, V. (2017, June). *An introduction to information security.* National Institute of Standards and Technology, 800-12. https://csrc.nist.gov/publications/detail/sp/800-12/rev-1/final

Office of the Press Secretary, White House. (2013, February 12). *Presidential policy directive—Critical infrastructure security and resilience* (PPD-21) [Press release]. https://obamawhitehouse.archives.gov/the-press-office/2013/02/12/presidential-policy-directive-critical-infrastructure-security-and-resil

Rose, S., Borchert, O., Mitchell, S., & Connelly, S. (2020, August). *Zero trust architecture.* National Institute of Standards and Technology, 800-207. https://csrc.nist.gov/publications/detail/sp/800-207/final

Ross, R., Pillitteri, V., Dempsey, K., Riddle, M., & Guissanie, G. (2020, February). *Protecting controlled unclassified information in nonfederal systems and organizations.* National Institute of Standards and Technology, 800-171. https://csrc.nist.gov/publications/detail/sp/800-171/rev-2/final

Sanger, D. E. (2012). *Confront and conceal: Obama's secret wars and surprising use of American power.* Crown.

Sanger, D. E. (2018). *The perfect weapon: War, sabotage, and fear in the cyber age.* Crown.

SANS Institute, Industrial Control Systems. (2016). *Analysis of the recent reports of attacks on U.S. infrastructure by Iranian actors.*

Scarfone, K., & Hoffman, P. (2009, September). *Guidelines on firewalls and firewall policy.* National Institute of Standards and Technology, 800-41. https://csrc.nist.gov/publications/detail/sp/800-41/rev-1/final

Stockton, P. N. (2021, August 27). *Strengthening the cyber resilience of America's water systems: Industry-led regulatory options.* American Water Works Association. https://www.awwa.org/Portals/0/AWWA/Government/STRENGTHENINGTHECYBERRESILIENCEOFAMERICASWATERSYSTEMS-INDUSTRY-LEDREGULATORYOPTIONS.pdf

Stouffer, K., Lightman, S., Pillitteri, V., Abrams, M., Hahn, A. (2015). *Guide to industrial control systems (ICS) security.* National Institute of Standards and Technology, 800-82. https://csrc.nist.gov/publications/detail/sp/800-82/rev-2/final

Strand, J., Asadoorian, P., Robish, E., & Donnelly, B. (2013). *Offensive countermeasures: The art of active defense.* PaulDotCom.

Trend Micro, Inc. & Organization of American States. (2015). *Report on cybersecurity and critical infrastructure in the Americas.* https://www.oas.org/es/sms/cicte/ciberseguridad/publicaciones/2015%20-%20OAS%20Trend%20Micro%20Report%20on%20Cybersecurity%20and%20CIP%20in%20the%20Americas.pdf

U.S. Department of Homeland Security. (2003). *National strategy for the physical protection of critical infrastructures and key assets.* National Infrastructure Advisory Council. https://georgewbush-whitehouse.archives.gov/pcipb/physical.html

U.S. Department of Homeland Security & Federal Bureau of Investigation. (2016, December 29). *Grizzly Steppe—Russian malicious cyber activity.* https://www.cisa.gov/uscert/sites/default/files/publications/JAR_16-20296A_GRIZZLY%20STEPPE-2016-1229.pdf

U.S. Department of Homeland Security & U.S. Environmental Protection Agency. (2015). *Water and wastewater systems sector-specific plan.* https://www.cisa.gov/sites/default/files/publications/nipp-ssp-water-2015-508.pdf

Wüest, C. (2010). *The risks of social networking.* Symantec Corporation.

Zetter, K. (2015). *Countdown to zero day: Stuxnet and the launch of the world's first digital weapon.* Crown.

9

Case Studies

1.0 CASE STUDY 1: INFORMATION TECHNOLOGY FOR INDUSTRIAL PRETREATMENT REPORTING

1.1 Introduction and Business Context

The City of Grand Rapids, Michigan, Environmental Services Department (ESD) serves 272,000 people in 11 customer communities with 56 lift stations and a 40-mgd water resource recovery facility, a Utility of the Future Today–designated facility. The ESD also oversees stormwater management; green infrastructure; and energy, lighting, and communication in the city. In October 2005, the U.S. Environmental Protection Agency (U.S. EPA) published the Cross-Media Electronic Reporting Rule (CROMERR). The rule provides the legal framework for electronic reporting and authorizes and facilitates electronic reporting for environmental programs while maintaining the level of corporate and individual responsibility and accountability that exists for paper submissions. In Grand Rapids, there are currently 84 permitted industrial users that use CROMERR. CROMERR requirements for electronic reporting are comprehensive and include

- timeliness of data generation,
- integrity of the electronic document,
- submission knowingly with intent and not by accident,
- opportunity to review and repudiate copy of record,
- acknowledgement of receipt, and
- determining the identity of the individual uniquely entitled to use a signature device.

ESD joined in the joint Water Research Foundation/Water Environment Federation research project Utility Analysis and Improvement Methodology (UAIM), which was designed to provide a systematic approach to evaluate business processes to identify and implement changes that will benefit the utility and its stakeholders. The CROMERR process was selected by ESD as the process to evaluate with UAIM.

The business driver/opportunity was twofold. First, industrial users are required to submit reports to the ESD's Industrial Pretreatment Program (IPP), and they were requesting electronic reporting capabilities. The initial process involved faxes and mailing paper copies of reports. The second driver was the city's desire to leverage technology to streamline the reporting process to reduce the time spent reviewing reports and to have electronic data storage capabilities to eliminate paper files.

1.2 Technology Description

There were no commercial off-the-shelf software systems that were CROMERR compliant at the time, so ESD partnered with its current pretreatment software provider to develop, create, and integrate an electronic reporting system software with existing pretreatment software.

The technology developed is a web browser–based electronic reporting system that allows users to submit secure electronic report data using designated data fields and that can be transferred to an existing pretreatment software database.

1.3 Stakeholder Involvement

This effort involved staff from a number of organizations including ESD; industrial users; U.S. EPA; the Michigan Department of Environment, Great Lakes, and Energy; and the software provider. Within ESD, the initial effort to define the project and champion the effort to obtain funding was led by the ESD manager. This also included negotiating the contract with the software provider, obtaining city commission approval, and coordinating with U.S. EPA contacts. Once the project was approved, contracted, and funded, it was led by the IPP supervisor.

The IPP supervisor led the interactions between internal and external stakeholders. This included soliciting input from IPP inspectors and industrial users regarding software functionality, coordinating with the software provider to develop the necessary functionality, and working with U.S. EPA to obtain CROMERR approval in coordination with the software provider. The IPP inspectors, as primary users of the system's outputs, were involved in providing feedback on software functionality across multiple iterations through deployment.

Industrial users were another key stakeholder group because the system could provide benefits to them in the form of reduced time and cost for regulatory reporting. They provided letters of support to the City of Grand Rapids city manager and provided feedback on design and function after initial training. The software provider consulted with one industry during iterations testing, who provided testing and feedback on design and functionality throughout the process.

The Michigan Department of Environment, Great Lakes, and Energy is the approval authority for Grand Rapids' IPP program. It was actively involved in discussions regarding CROMERR and what items were needed for the IPP program, including procedure manual changes, ordinance changes for legal authority, and program modification submittals.

U.S. EPA is the approval authority for the compliance of the CROMERR program and its submission outputs with CROMERR requirements. They provided a contact for program questions, and U.S. EPA information technology (IT) contractors worked with the software provider team and Grand Rapids to complete the CROMERR checklist to submit for CROMERR approval.

The software itself was developed and delivered by a commercial pretreatment software provider. The software provider's project manager was responsible for coordinating the development of a CROMERR-compliant system through meetings with the IPP supervisor, U.S. EPA IT contractors, and their in-house software development team. The project manager coordinated closely with the IPP supervisor to provide status updates, design mock-ups, software functionality components, and testing throughout the entire process.

The project manager for the software provider and the IPP supervisor had been coordinating on other pretreatment software since the early 2000s and continued to coordinate with this CROMERR project. This allowed for a familiarity with design approach and how to present feedback and/or concerns constructively in an efficient manner. The roles were clear from the beginning of the project and the expectation of both the software provider and the City of Grand Rapids was to develop and implement a CROMERR-compliant electronic reporting system. There was concern of shifting expectations when the software was purchased; however, meetings held with the software provider and the new ownership verified that the expected goal had remained unchanged.

1.4 Role of Data

Data collection and reporting were the focus of this project, with the goal of streamlining the process to reduce the time and cost for both the data providers and the data users. The data sources included the following:

- Industrial user wastewater sampling data from laboratory analysis reports. Data are hand entered by industrial users into the software or can be imported following a standard file format. The data file can be generated by the industry's laboratory information management system.
- Data about each industry and its sampling and reporting requirements were gathered into a commercial off-the-shelf industrial pretreatment software database provided by the software vendor contracted by ESD to create the CROMERR-approved electronic reporting solution.

With data as the primary concern of this effort, there were a number of challenges to address:

1. Identifying the data fields for electronic entry. Unlike the paper forms that had been used in the past, electronic data entry requires the definition of rules to ensure data quality. Data from paper copy reporting forms were used to determine necessary data fields, and testing was done to import industrial user contract lab data into the reporting system for transfer to the pretreatment software database. System developers and Grand Rapids Pretreatment Program staff worked to determine the most efficient layout of the data fields, what data were required, and which data would be optional.
2. The commercial off-the-shelf industrial pretreatment software database used by ESD was a client/server software application, but the electronic reporting solution was being built as a web browser–based software as a service application and database. Challenges were encountered aligning and synchronizing those databases. This was ultimately overcome by co-locating the databases and building a data synchronization process.
3. To meet CROMERRR approval, industrial users who submit reports using the software must have their identity validated. Identify validation was performed by IPP inspectors collecting signature agreements as part of scheduled site visits.
4. CROMERR requirements were followed to ensure the validity and security of electronically collected data. Information technology staff followed the U.S. EPA checklist to ensure that these requirements were met. An important element of electronic reporting is that industrial users who are submitting reports through the system create an electronic signature. The electronic signature must be bound to the report being submitted in such a way that attempting to tamper with the contents of the electronic report would invalidate the report. The software is required to prove that the report has not be tampered with. To meet this requirement, the software vendor created a registration process for the industrial user that followed best practices to create a username and a strong password and establish five knowledge-based questions that are used together to create an electronic signature for that person. These are hashed using SHA-256 standards. At the time of report submission, the user must enter their password and answer a knowledge-based question correctly. If correct, the contents of the report, including every data element, attached files, and the electronic

signature, are combined together into a compressed .zip file called a copy of record (COR). The .zip file is encrypted, and a globally unique hash of the file is created. The hash is shared with the user as proof of submission. At any time, Grand Rapids IPP staff can validate that the electronic report stored in the system has not been tampered with through a "validate COR" process that performs the same hashing process and confirms that the hash generated matches that of the originally submitted report. Sources of data came from user-entered data from laboratory reports or an import feature to provide the input data to the software.

As the logistics of the system architecture defining where the data are stored were being worked out, the issue of data ownership came up. It was important to recognize that ESD retains ownership of the data, even if they are stored in the cloud on an outside vendor's system. Data ownership and use requirements should be addressed in the contracts with the software vendor.

1.5 Timeline, Cost, and Approach

The proposed timeline was 116 weeks (2.2 years) to develop the software in three phases with an approved budget of $300,000. This was determined with many assumptions for the development process, and the timeline was extended as the project began as a result of a better understanding of developing criteria and increased functionality design throughout the process. The software functionality was tested on a small group of industries for feedback before a tiered software rollout plan. During this time, the software provider was acquired multiple times by larger companies. This resulted in leadership changes of the development team and dynamic priorities.

1.6 Outcomes

The implementation of the IT capabilities for industrial pretreatment reporting was complemented by process changes that were identified as part of the UAIM project. IT-enabled process improvements resulted in a notable reduction in people hours from ESD staff to review the reports, and a reduction in time for the industry's reporting time. Another outcome of using a digital approach to reporting is the ability to generate email notifications on a real-time or daily basis depending on the report. The utility can define who receives each notification report. Examples of these reports include

- a list of reports that have been electronically submitted and ready for review that is in real time; this notification report also indicates any account lockouts and forgotten password notifications;

- a report of data import failures including the specific error and where it is located in the data file;
- a report of permit limit violations by industrial user and parameter, which reduces the time for IPP staff to review and interpret paper reports;
- a list of samples and results imported by industrial user; and
- a report that indicates if the samples imported are from authority sampling or from industrial user sampling.

These electronic reports that analyze and organize the data reported by the industry users allow IPP staff to do a quick data review without having to open the software and navigate to each section to find the data.

1.7 Lessons Learned

Electronic reporting does improve the efficiency of data collection and responding to the data, but it does not eliminate all data quality issues. Users still make simple mistakes that require the report to be revised and resubmitted. Items such as missing flow values, inputting incorrect sample types and sample start and end dates, forgetting attachments, adding wrong attachments, missing samples, selecting the wrong samples, entering incorrect sample results, missing pollutants, entering sample result data that does not accurately correspond to the lab's analytical report, incorrect method numbers, and using wrong report types are all examples of data issues that have occurred.

Designing a user interface that will reduce manual data entry for users is key to preventing errors, such as additional drop-down boxes and additional warnings to users for data inputs that do not align with data fields, and warnings for incorrect inputs, would be beneficial, and are currently being developed to alleviate or reduce the known user input issues.

Additional lessons learned included the following:

- Keep the design and the workflow as simple as possible while still meeting the needs of the reporting requirements. This makes it easier for users and improves the usability of the tool.
- ESD owns the data even when it is stored on outside servers.
- Development software frameworks and standards change, and the initial framework is no longer supported. This required a change for the development team, resulting in additional time to complete the project.

- New personnel need to be trained, and someone at ESD needs to have administrative rights to manage the users and ensure that everyone has the system access appropriate for their role.
- Different software browsers behave differently with applications. At first, the software worked on all browsers except Safari, but this issue was resolved quickly.

2.0 CASE STUDY 2: INFORMATION TECHNOLOGY FOR ASSET MANAGEMENT

2.1 Introduction and Business Context

Hampton Roads Sanitation District (HRSD) is a political subdivision of the Commonwealth of Virginia that was created in 1940 and serves 1.7 million people across 20 counties and cities of southeast Virginia and the eastern shore. It operates more than 100 pumping stations, nine major treatment facilities, and eight smaller facilities with a combined treatment capacity of 249 million gallons per day. Like many utilities in the eastern United States, HRSD was challenged by the effect of wet weather events on its sanitary sewer system and entered into a consent decree with Virginia Department of Environmental Quality in 2007.

The efforts to address the consent decree requirements included condition assessment and involved working with the utility's Geographic Information Systems (GIS) Division to deploy mobile devices to collect data. To complement these efforts, a computerized maintenance management system (CMMS) from Infor was implemented to track asset condition and maintenance records.

Hampton Roads Sanitation District's asset management journey over the last 15 years, as presented in Figure 9.1, is an example of how the application of IT can evolve over time to support continuous improvement. Starting with the basics of asset identification and building on that foundation with additional IT capabilities including the integration of GIS and CMMS, as well as organizational and process changes, has enabled HRSD to progressively enhance the value of its IT investments over time.

In 2015, HRSD reorganized and created an Asset Management Division. The driver for the creation of this program was work centers' frustration that they could not find assets, that it was a challenge to assign work orders, and that they were not able to properly evaluate the amount of time spent on the program to demonstrate the need for additional resources. Experience within HRSD as well as knowledge sharing across the industry created a more in-depth understanding of asset management best practices and the tools and

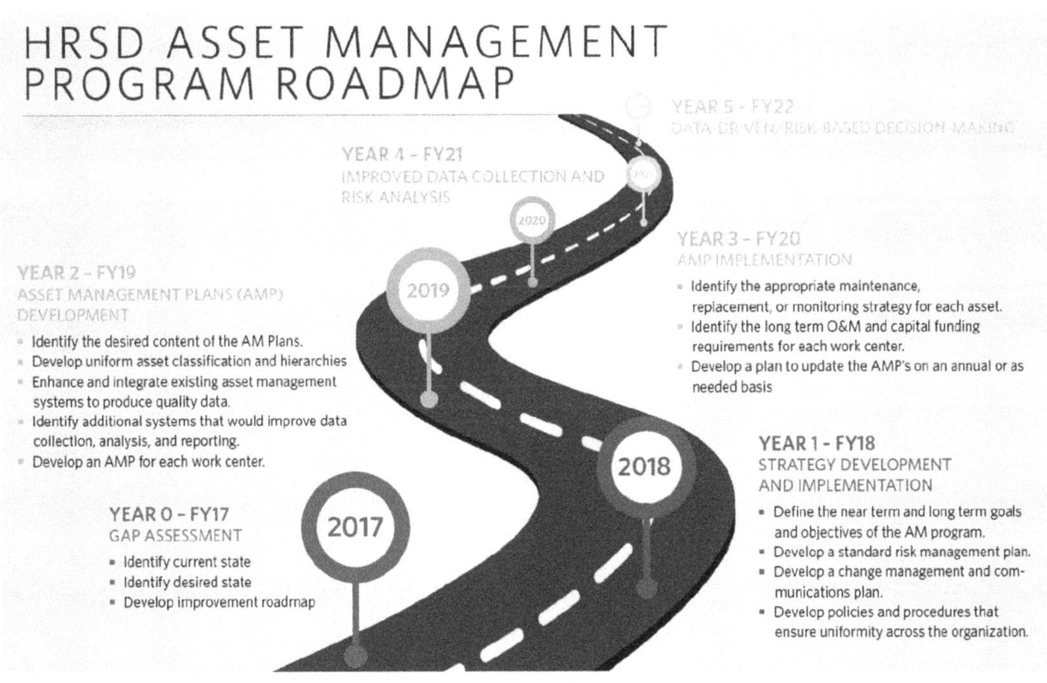

FIGURE 9.1 HRSD Asset Management Timeline (Reprinted with permission from Anas Malkawi, Hampton Roads Sanitation District)

processes needed to maximize the value of the program beyond consent decree compliance. The group responsible for capital improvement planning was requesting data-driven, risk-based decision making, so HRSD initiated an assessment to take its asset management program to the next level.

In 2016, HRSD decided to adopt the International Standards Organization (ISO) 550001 framework. Figure 9.2 presents HRSD's application of this framework. "The ISO 55001 management system provides a framework to establish asset management policies, objectives, processes and governance, and facilitates an organization's achievement of its strategic goals. ISO 55001 utilizes a structured, effective, and efficient process that drives continual improvement and ongoing value creation by managing asset-related cost, performance and risk" (International Standards Organization, n.d.). This framework has provided a valuable process for identifying priorities and taking action to improve asset management practices and business outcomes, and to provide the context for the use of IT tools.

2.2 Technology Description

The technologies used by HRSD have evolved over time. The program began with GIS to spatially locate assets and the CMMS to manage maintenance

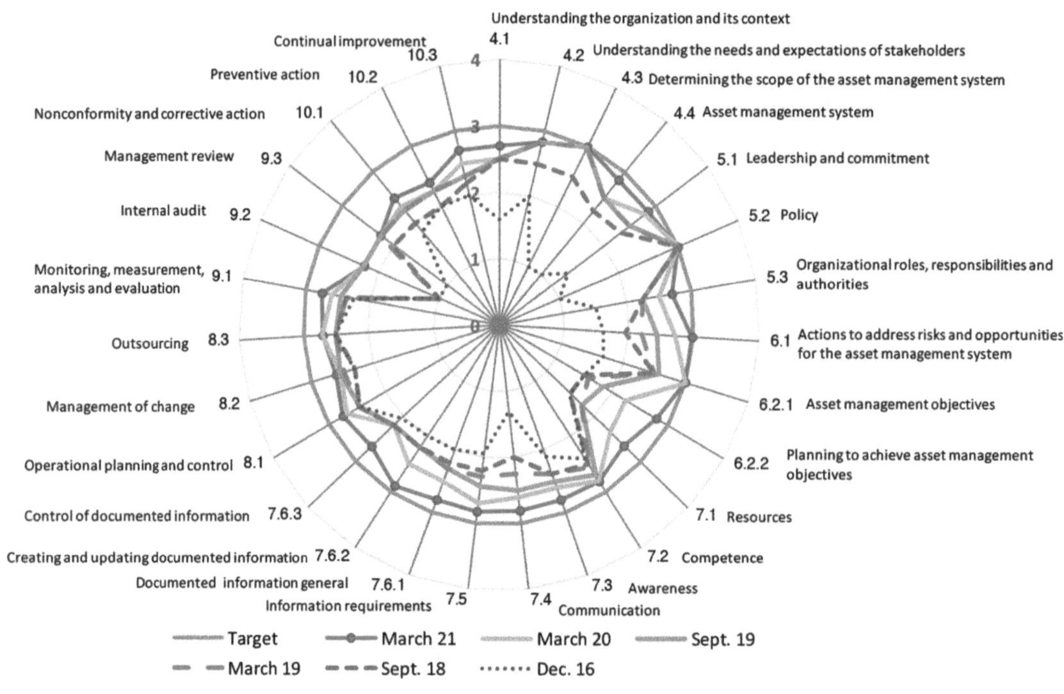

FIGURE 9.2 ISO 55001 Framework (Reprinted with permission from Anas Malkawi, Hampton Roads Sanitation District)

activities. Improvement needs were identified as staff gained experience working with the systems, which resulted in a reevaluation of the CMMS. Through this effort, it was determined that the CMMS itself was not a problem, and that by modifying business processes, as well as improving data and system integration capabilities, greater value could be achieved.

In 2016, following the creation of the Asset Management Division, HRSD conducted a visioning workshop. This effort illuminated the critical role of data in the overall asset management program and led to a data needs assessment. This assessment provided the foundation for future IT system recommendations.

It is important to recognize that as HRSD's efforts evolved, the focus was not on the IT systems themselves, but instead on the processes and data needed to support business activities. It is this approach that resulted in the identification of the gap that vertical assets did not have a good visualization tool. Building information modeling (BIM) had been in use for more than 10 years for design, but the data had not carried forward beyond design. Closing that gap was set as a priority. It was noted that standard technology for utilities was not in line with the vision HRSD had, so it began exploring the practices and tools in use at other types of organizations including transportation departments and universities. Hampton Roads Sanitation District collaborated with ESRI and Autodesk on 3-D representation of vertical data and realized the need to start building the data sets.

A BIM pilot was part of a strategic capital program, the Sustainable Water Initiative for Tomorrow (SWIFT) demonstration facility that was being built. BIM requirements were created for this project to include design and construction standards that could be used to improve as-built data. Figure 9.3 presents an image of the use of BIM for vertical asset management.

2.3 Stakeholder Involvement

The asset management program at HRSD involved stakeholders across the organization. It was realized early on that to build a successful program, HRSD needed to address people and business processes. The program started with the maintenance staff who were on the front lines of data collection and who had the responsibility to manage the assets that were part of the consent decree. Workshops were the primary mechanism for identifying and discussing needs. As the efforts expanded, the approach expanded to include both a bottom-up and top-down approach.

A steering team was established to demonstrate executive support for the program and the changes in processes and expectations that were required.

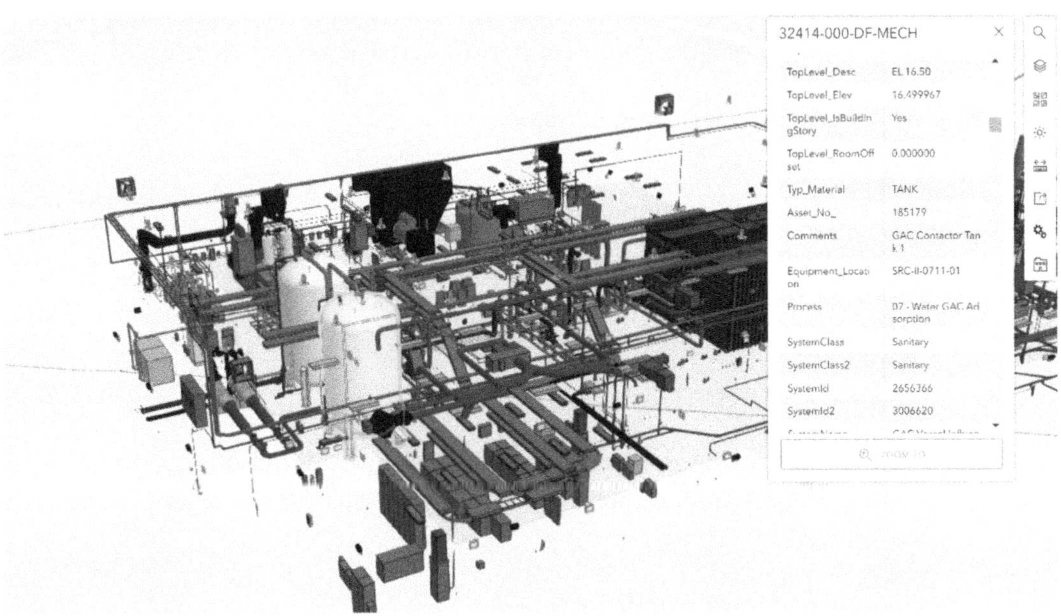

FIGURE 9.3 Vertical Asset Management BIM Model (Reprinted with permission from Anas Malkawi, Hampton Roads Sanitation District)

This team also provided broader business context for decision making. For example, maintenance staff were focused on meeting specific system performance standards "no matter what." The steering team, however, could evaluate the cost of meeting the standard against the trade-offs and opportunity costs for the utility's operations as a whole.

The primary stakeholders of the program were those staff involved in managing and maintaining assets and making business decisions using those assets. Senior IT staff are part of the steering team and actively involved in developing the business processes and ensuring that the technology capabilities were in line with the new processes and procedures.

All stakeholder needs were considered by addressed training needs, defining staff competencies, supporting a culture of continuous improvement, and communicating the purpose of changes and the benefits to each group.

2.4 Role of Data

As mentioned earlier, data were the primary focus of the asset management program. A key finding from the visioning workshop was the need to determine the data required to support asset management decisions. In 2018, a data needs assessment was conducted that identified primary data sets, gathered information about the data sets, and identified data gaps for each asset category. This information was used to prioritize data collection and improvement efforts and align requirements for data to inform performance measures, risk assessments, and asset replacement planning.

The asset groups that were part of this assessment were

- gravity sewers and manholes
- force mains
- wastewater pumping stations
- wastewater treatment facilities
- facilities
- fleet
- master metering

The following data types, which are considered important to asset management planning, were considered for each of the asset groups/asset categories:

- ID and physical description
- risk analysis (risk and consequence of failure)
- service level/performance
- condition assessment

- maintenance/work management
- operations and maintenance costs
- financial

The data sets evaluated varied depending on the asset group/asset category being considered. However, for each data set evaluated, the following information was sought:

- Does the data set exist? (Y/N)
- What is the source of data?
- Who is responsible for the data?
- Which information system is used to store and maintain the data?
- How complete are the data? (How much data is in the system of record?)
- How important are the data for asset management (e.g., essential, important, useful, not required)?
- A data maturity score (DMS) was assigned to each data set on a scale of 0 to 5 with 0 meaning no data are maintained and 5 meaning leading practice is applied.

The result of this data assessment was a set of recommended data management improvement actions. The IT systems were framed as the tools to support data collection, management, and analysis, rather than being the primary focus of the program.

2.5 Timeline, Cost, and Approach

Hampton Roads Sanitation District's journey began when the utility was founded because its culture has always been focused on continuous improvement. The asset management–specific effort began in 2007/2008 with the consent decree and became even more focused in 2015 when the Asset Management Division was created. There is no specific project or budget for this effort, but it is instead a collection of efforts across all aspects of the organization to align data, tools, processes, and people with business outcomes.

2.6 Outcomes

There have been three primary outcomes of the asset management program:

1. Data quality has improved. The systematic approach to evaluating asset management data needs and prioritizing activities to eliminate data gaps had a direct measurable effect.

2. HRSD has a risk-based capital program. By improving data quality and focusing on processes, the business can now use the data and decision-making frameworks to prioritize activities and investments each year.
3. Progress is being made to achieve higher levels of maturity on the ISO 55001 framework, and HRSD is close to achieving its initial goal of competency on all aspects. Future activities will be planned based upon the business value of additional levels of maturity.

2.7 Lessons Learned

There are two primary lessons from HRSD's asset management efforts. The first is that it is not the IT aspects of effort that are difficult; it is working with people, what engineers often refer to as the "soft" side. The soft side is indeed the hard part. Each stakeholder group—in fact, each person—has a different perspective that has been formed based on their past experiences and the expectations placed on them. Each of these different perspectives can be valid. The leaders need to make a decision based upon what makes the best sense at the time and then work to help people expand their perspective to understand why a decision was made. It was useful for HRSD to help staff understand that "we are all asset managers."

The second lesson learned was the value of the data needs assessment. This was used to improve data quality in areas based on prioritized needs and resulted in the beneficial outcomes described in Section 2.6. Focusing on data, rather than systems, enabled a more productive set of actions than focusing on IT.

3.0 CASE STUDY 3: CYBERSECURITY

3.1 Dam Control Vulnerability: Stuxnet—A Game Changer

During the second Bush administration, there was great concern about the rapid progress of the Iranian enrichment program, which would likely lead to the attainment of weapons-grade uranium. The center of that operation was, and is, the Natanz Laboratory located in the middle of the desert about 33 km from civilization. The facility, technically known as a "fuel enrichment plant," is one of 17 other Iranian nuclear facilities. The facility uses centrifuges to concentrate and separate U-235 from uranium hexafluoride gas; the facility was planned to operate approximately 19 000 centrifuges. Natanz is "air-gapped," in that it is isolated from the outside world both physically and electronically; it is essentially an island. It is heavily fortified, the processing facility being underground, covered by 22 m of earth. The facility was designed to be impregnable, both physically and electronically.

Code-named "Olympic Games," an effort was begun to avoid a direct conflict with Iran; Israel saw the effort by Iran as an existential threat and was preparing large-scale air strikes, which would likely have precipitated a ground war and major regional escalation. Out of this operation came the malware known as "Stuxnet" (a combination of keywords .stub and mrxnet.sys). Stuxnet, technically a worm, is the first known offensive cyberweapon designed specifically to inflict damage equipment in the real world—it was truly a game changer.

Stuxnet takes advantage of what are termed "zero-day" vulnerabilities. Zero days are unknown vulnerabilities in software that threat actors use to infiltrate and exploit unpatched operating systems. Stuxnet used four zero-day vulnerabilities to infect the Windows operating system. The worm installed rootkits, allowing complete control of the operation; Stuxnet is the first known use of programmable logic controller (PLC) rootkits. Command and control of the worm was done through two websites located in Denmark and Malaysia, though these were not used after the initial stages of the operation. The worm also used stolen digital certificates for many of its drivers to allow it to appear legitimate.

Initially, the worm was designed to close the discharge valves of the centrifuges to create overpressurization and waste gas—the centrifuges operate in a vacuum, and the gas solidifies at low pressures—instantly destroying the centrifuge. This method proved less than efficacious; the Iranians found a workaround and damage was limited. This version of the worm was updated on site several times as more operational data were observed by the mole and reported back to the consortium. At this point, the mole lost access to the facility for reasons unknown; it is also possible the mole did not need access any longer. Concurrent to the on-site injection operations, several Iranian contractors performing work at the facility were compromised or their computers infected with the second version of the worm. It is likely the newer version of the worm was delivered by these unwitting employees using the planted USB drive method; however, it would have been easier to infect the contractor's internal networks. The worm does not attack computers; rather, it was designed to attack Siemens supervisory control and data acquisition (SCADA) software and PLCs

3.1.1 The Attack

Stuxnet is a sophisticated piece of software. It is estimated that a multinational team of coders took up to 3 years to develop the worm. Key to the attack was the ability of the worm to monitor and record normal operating data; the operator saw what looked like normal operating parameters on the human–machine interface screen while the centrifuge was operating adversely.

The worm was designed to infect the PCS 7 SCADA software files and compromise Siemens S7 PLCs that controlled the rotational speed of the centrifuges. The centrifuges normally spin at a rate of 63 000 rpm; because of manufacturing defects, however, the Iranians ran the centrifuges at about 4000 rpm lower to avoid cracking of the rotors. The actual attack was silent and autonomous. While the operators observed normal operation on their screens, the worm brought the centrifuges almost to a halt and then ran them rapidly up through the critical intermediate speed of 59 000 rpm to 40% over normal operating speeds (84 000 rpm). This alternation between low and overspeed conditions created significant vibration in the rotor and rotor bearings, which essentially destroyed themselves after a few cycles. This method effectively destroyed more than 1200 centrifuges but did not significantly delay the enrichment program.

3.1.2 The Counterattack

As one would imagine, the Iranians were not pleased and quickly dissected the code, determining with reasonable certainty that the United States and its allies were behind the cyberattack. Following discovery of the Stuxnet attack and attribution in the technical press, Iran embarked on an aggressive counterattack involving businesses and critical infrastructure in both countries.

The Arthur Bowman Dam in Oregon is 245 feet high and 800 feet wide, impounding 233 000 acre-feet of water for irrigation purposes. The Iranians targeted this dam in 2013 as one of their responses to the Natanz attack. Their reconnaissance used Google; there are several search techniques and syntax that are called "Google dorking," and they turn Google from a simple search engine into a powerful research tool. The next tool in their arsenal was Shodan, a specialized search engine that seeks out industrial control systems connected to the internet.

What the Iranians found was the Bowman Avenue Dam in Rye Brook, New York. The Bowman Avenue Dam is 20 feet high and 50 feet wide, impounding the flood stage of the Blind Brook. The Iranian team found an unsecured wireless modem that would have been used to control the dam's slide gate remotely (it was not connected to the gate's control system). It is speculated that this was the attack vector for the Arthur Bowman Dam; the dam gates would have been opened or closed to cause flooding or overtopping—either would have been a problem, though loss of life was not likely. In their assessment of the data they acquired, they neglected to do on-the-ground reconnaissance; this led to a high-profile failure for them, but it was a wake-up call for the rest of us.

3.2 The "NotPetya" Attack: A Lesson in Global Vulnerability

In 2017, one of the most widespread and devastating cyberattacks was perpetrated against the worldwide shipping giant Maersk. It started on a quiet afternoon in June, when staffers began seeing messages advising them that their file systems were being repaired, while others got the message that their important files had been encrypted. Maersk is a global shipping titan, responsible for 76 ports around the globe and more than 800 vessels, carrying all manner of goods and constituting about a fifth of global trade.

Beginning in 2012, Ukraine and Russia had been slugging it out in an undeclared war that served as a proving ground for Russia's cyberwarfare tactics. A group of Russian hackers called Sandworm had thoroughly compromised the Ukrainian government and dozens of Ukrainian companies. One way the Russians were able to apply such a broad and sweeping campaign of destruction was through the compromise of the Linkos Group, a small software firm that markets an accounting software package called M.E.Doc; this software was used by nearly everyone who did business in Ukraine and gave Sandworm a vast attack surface to work with. Sandworm had hijacked the firm's update servers early in 2017, and this gave them backdoor access to the thousands of computers running M.E.Doc.

That June, Sandworm released a particularly vicious cyberweapon called "NotPetya," which spread rapidly and automatically. The code was indiscriminate in who it attacked; it was designed to do the most damage as quickly as possible and with the widest possible swath of destruction. NotPetya was composed of two major elements: a penetration tool called EternalBlue, created by the U.S. National Security Agency and leaked in early 2017, and Mimikatz, a software application that had the ability to pull user passwords out of RAM and reuse them to compromise the target machine. Whereas Microsoft had issued a patch for EternalBlue, Mimikatz allowed retrieval of passwords, which allowed those passwords to infect unpatched machines anywhere in the world. NotPetya was not actually ransomware—its intent was purely destructive. There was no decryption key for the destroyed data.

Within hours of NotPetya's release, it had raced around the world and infected countless computers. FedEx's European subsidiary, TNT Express; several French companies; a hospital in Pennsylvania; the pharmaceutical company Merck; and, of course, Maersk were infected. The radiation monitoring system at the Chernobyl nuclear facility went offline. The infection even spread back to Russia, infecting state oil company Rosneft. The result was damages of about $10 billion.

In July 2017, Ukraine's cybercrime unit seized servers from Intellect Services, the company that produced the M.E.Doc software. Analysis of the servers showed that they had not been updated for at least 4 years, and security patches were nonexistent. There was evidence of Russian presence in the servers, and several employees' accounts had been compromised.

An incident response team was assembled, and an emergency recovery center was put together in Great Britain to mitigate and recover from the NotPetya attack. This was a global effort and required hundreds of staffers working 24/7 to rebuild the network. All computer equipment had been confiscated, and new computers had been obtained and then distributed to recovery personnel. Staff began rebuilding servers from the ground up. However, this effort came to a grinding halt when it was realized that there was no clean backup of the company's domain controllers.

This is an effective and decentralized backup strategy that ordinarily would have allowed quick recovery from a localized event; however, no one had visualized a scenario in which all the company's domain controllers were wiped out in a massive attack. Maersk staffers finally found one pristine backup in their Ghana office. By a stroke of luck, a blackout had knocked the server offline before the NotPetya attack and had disconnected it from the network. It contained the single clean copy of the company's domain controller data, and discovery of it was a source of great relief to the recovery team.

The recovery team began bringing up Maersk's core services, concentrating on port services. Key to this was the ability to read a ship's inventory—each ship had 18 000 containers—and determine what was where and where it was bound for.

When it was over, Maersk estimated that NotPetya had cost the company between $250 million and $300 million, though many believe this number was lower than actual. Costs down the line were also significant; trucking companies lost tens of millions of dollars, TNT Express lost about $400 million, while Merck lost a staggering $870 million. The disruption to the global supply chain, of which Maersk is a major component, was extensive, and losses accumulated into the billions of dollars.

3.3 The Disgruntled Employee: The Maroochy Shire Incident

In 2001, an insider attack on the Maroochy Shire Water District in Queensland, Australia, resulted in significant environmental and economic damage. Vitek Boden formerly worked for Hunter Watertech, an Australian firm that installed SCADA radio-controlled wastewater equipment for the Maroochy Shire Council in Queensland, an area of great natural and scenic beauty. Boden applied for a job with the Maroochy Shire Council, but the council decided not to hire him. Almost immediately, the system began to experience a series of faults. It was observed that pumps were not running when they should have been, alarms were not reporting to the central control room, and at various times, communications were lost between the central control room and the remote pumping stations. The wastewater collection and treatment system processed 9 million gallons of wastewater a day, using 142 pumping stations placed about the shire.

The shire hired experts to look into the situation. They did a thorough analysis of the existing control systems and eliminated many of the common problems that might have occurred, such as bad code or hardware issues. This left them stumped because as soon as it was determined that a pump was malfunctioning, teams were sent to that pumping station, only to find that nothing was wrong with the equipment, and everything appeared to operate normally. Data traffic was monitored and examined between the central control room and the pumping stations, and between the pumping stations. Upon issuing a command to change a control parameter, an engineer was surprised to find that his change had been canceled and the original parameter was still in the configuration. Upon further examination, it was determined that one pumping station had issued the reset command and changed the configuration data. This pointed directly to a human operator.

The engineer decided to set a trap. He changed the identification number of the suspected pumping station and observed that control commands were still coming from the same pumping station under the old identification number. Someone had hacked the communication system and was pretending to be the pumping station control system. Boden was immediately suspected. The shire had hired a detective firm to track Boden when a series of errors hit several pumping stations. A detective trailing Boden noticed his car not far from one of the pumping stations and called the police. A chase ensued, and Boden was caught and arrested. A laptop with pirated SCADA system software was found along with a radio transceiver.

Boden had decided to get even with both the council and his former employer. He equipped his car with stolen radio equipment attached to the laptop computer. Between February and April 2000, on at least 46 occasions, he issued radio commands to the wastewater equipment he had installed while working for Hunter Watertech. Boden caused 200 000 gallons of raw wastewater to spill out into local parks, rivers, and even the grounds of a Hyatt Regency hotel. It was reported that marine life died, the creek water turned black, and the stench was unbearable for residents. Boden was sentenced to 2 years in jail and ordered to reimburse the council for cleanup. It was thought that Boden concocted a scheme to get his job back to fix the errors he was causing. The Maroochy incident was an early wake-up call to those in the IT security field. The SCADA system was not designed with security in mind. Boden's attack became the first widely known example of someone maliciously breaking into a control system.

3.4 New Awareness in the United States: The Oldsmar, Florida, Hack 2021

In February 2021, a yet-to-be-identified hacker was able to breach the water treatment facility in Oldsmar, Florida, near Tampa. This incident marks the first time that a water treatment facility was attacked in this manner; previous attacks involved unsuccessful ransomware attacks: Onslow Water and Sewer, Jacksonville, North Carolina, 2018; Fort Collins, Colorado, 2019; and Riviera Beach, Florida, 2019. Other attacks involved illegal activities by employees or confirmed breaches by unidentified threat actors that did not result in damage or loss and were contained. Oldsmar is the first known breach that specifically targeted process control systems and successfully manipulated processes for nefarious purpose.

The attacker changed the setpoints in the control system to add 11 100 ppm of sodium hydroxide instead of the 100 ppm normally added. That the attacker was able to breach the system and then access facility control systems to the extent that operating parameters could be adjusted is unprecedented in a domestic water treatment facility. The attacker breached the system twice on February 5, using an unsecure remote viewing software package called TeamViewer. Fortunately, an operator discovered the change and reversed it; however, the physical characteristics of the system would not have allowed the dosage to take effect immediately, which provided some time for the operators to respond.

Several factors contributed to the breach. It has been disclosed that on the day of the attack, a computer on the Oldsmar facility network visited a site that infected the computer with malicious code. This is called a "watering hole attack" and uses commonly visited websites to lure unsuspecting victims into clicking a link, which in turn creates an exploitable vulnerability. This is apparently what happened in this case. The threat actor was able to access the facility through compromised data that were found on the website and allowed the attacker to access TeamViewer, likely with misappropriated credentials. It has been determined that the attacker was able to gather in excess of 100 separate pieces of data about the facility systems, including operating systems and CPU types, browsers in use, data inputs and outputs, and details about process control software. Investigators followed the trail to a website on the "dark web" that sells access credentials.

Using the illicit credentials and unsecure software, the attacker was able to easily access the Oldsmar network. Among the major factors contributing to the breach was poor password policy; the same password was used for all systems within the facility. The facility had an interconnected office and process control network, which allowed both to be compromised after visiting the website. There were no firewalls in use. Finally, computers at the facility were still running Windows 7 and were likely unpatched. The threat actor likely accessed the system by exploiting cybersecurity weaknesses and the apparent lack of cyberhygiene.

This attack was wholly facilitated by human error. This was compounded by a failure to upgrade and update computer equipment. Education also played a part; the staffer who accessed the watering hole was not trained in cyberhygiene or would have avoided the website that aided the exploit by stealing data from the staffer. This is not an uncommon scenario. This incident illustrates the vulnerability that can be exploited by targeting untrained staff and underfunded facilities. Just as pumps and pipes and fixtures must be replaced and upgraded, so must facility security be updated. The facility has since remedied these issues and does not allow remote access to the facility.

3.5 Infrastructure Billing and Customer Service in the Colonial Pipeline Ransomware Attack 2021

On May 8, 2021, the Colonial Pipeline Company announced it was shutting down operations because of a ransomware attack. This attack disrupted supplies of gasoline and other refinery products in the eastern United States, particularly in the Southeast. Colonial was attacked by a ransomware organization called Darkside, and Darkside's primary target was Colonial's billing and customer service communication systems. Colonial shut down all operations out an abundance of caution and to give the company an opportunity to analyze and mitigate the attack.

The Colonial Pipeline system is the largest such entity in the United States. Colonial is one of six large pipeline systems operating in the United States and moves about 2.5 million barrels per day (MBD) of refined petroleum products through 5500 miles of pipelines that stretch from the Gulf Coast to terminals throughout the Atlantic Coast. This area consumes roughly 6 MBD of petroleum products. The pipeline is vital to the economic health of the regions served, and the shutdown caused widespread panic and hoarding of fuel, particularly gasoline, which some consumers collected in plastic bags.

Darkside is an "entrepreneurial" ransomware hacking group based out of eastern Europe. The group appeared in August 2020 and announced that it would not target hospitals, schools, or "businesses that couldn't afford to pay." Its business model was as a "ransomware as a service," in effect licensing its ransomware and distribution infrastructure to criminal enterprises and taking a cut of whatever is realized from an attack. Darkside has tried to burnish its reputation by making a series of charitable donations. It claims to be apolitical and is "only in it for the money" and not to make problems for society. Darkside almost exclusively targets English-speaking organizations, and the malware is designed to shut down if it detects languages consistent with Russia's sphere of influence. Darkside ransomware has earned the organization more than $90 million in Bitcoin.

Colonial announced that it had paid the $5 million ransom (actually $4.4 million using cryptocurrencies) and was sent decryption software to restore its hijacked data. Because of the slow speed of decryption and the prudent verification of the decrypted data, it took about a week for Colonial to restart operations after a 6-day shutdown. Because of the size of the pipeline system, it took up to 2 weeks for product to arrive at terminals again (the products move at between 3 and 5 mph). A gallon of fuel sent into the system in Houston will take 2.5 weeks to reach New York.

The method of the attack and how the criminals gained access to the Colonial system is still under investigation and many details are confidential. It is theorized that Darkside could have bought access credentials to remote desktop software such as TeamViewer and Microsoft Remote Desktop from online markets. Another way to obtain credentials or to find login portals to obscure internet computer systems is to use a search engine like Shodan, which specializes in seeking out internet-connected control systems. One other theory is that login credentials were found in a deactivated account. It is known that a legacy virtual private network (VPN) was used to perpetrate the breach and install the malware. The VPN used single-factor authentication, which is, simply put, poor cyberhygiene.

Although Colonial has taken steps to harden its cyber defenses, the company readily admits that vulnerabilities still exist that can be exploited by a determined attacker. Though seizures are rare, the government was able to recover 60 of the 75 Bitcoin paid.

4.0 REFERENCE

International Standards Organization. (n.d.) *About.* https://committee.iso.org/home/tc251

Index

CPSIA information can be obtained
at www.ICGtesting.com
Printed in the USA
LVHW060604270522
719764LV00001B/1